Computing Projects

DIFFERENTIAL EQUATIONS

COMPUTING AND MODELING

C.H. Edwards Jr. △ David E. Penney
The University of Georgia

PRENTICE HALL Upper Saddle River, NJ 07458

Production Editor: *Carole Suraci*
Production Supervisor: *Joan Eurell*
Acquisitions Editor: *George Lobell*
Supplement Acquisitions Editor: Audra Walsh
Production Coordinator: *Ben Smith*

Simon & Schuster / A Viacom Company
Upper Saddle River, New Jersey 07458

Printed in the United States of America

10 9 8 7 6 5 4 3

ISBN: 0-13-504465-0

Prentice-Hall International (UK) Limited, *London*
Prentice-Hall of Australia Pty. Limited, *Sydney*
Prentice-Hall Canada Inc., *Toronto*
Prentice-Hall Hispanoamericana, S.A., *Mexico*
Prentice-Hall of India Private Limited, *New Delhi*
Prentice-Hall of Japan, Inc., *Tokyo*
Simon & Schuster Asia Pte. Ltd., *Singapore*
Editora Prentice-Hall do Brasil, Ltda., *Rio de Janeiro*

Contents

First-Order Differential Equations

Mathematical Models and Numerical Methods

Linear Equations of Higher Order

Introduction to Systems of Differential Equations

Linear Systems of Differential Equations

Nonlinear Differential Equations and Phenomena

Laplace Transform Methods

Power Series Methods

Fourier Series Methods

Eigenvalues and Boundary Value Problems

Preface

There is wide interest in the idea of including a computer lab component in the introductory differential equations course. This is a computing projects manual written to accompany Edwards & Penney, ***DIFFERENTIAL EQUATIONS with Computing and Modeling***, Prentice Hall, (1996). It contains expanded versions of the textbook's approximately 45 projects that have been augmented with detailed discussion and illustration of applicable *Maple*, *Mathematica*, and MATLAB syntax and techniques.

Each project typically begins with discussion of the investigation at hand in general terms — without reference to specific software. This opening discussion is then followed by separate **Using *Maple***, **Using *Mathematica***, and **Using MATLAB** sections that can be read independently. Thus we offer three separate computing threads that can be followed through the manual, depending on the computing environment chosen for use. However, it is common experience that understanding of a computing algorithm is enhanced by examination of its implementation in different environments (just as understanding of a standard mathematical topic may be deepened by study of complementary expositions). Even in the typical situation where a single computing system is made available for class use, we believe that many students will profit from occasional reading of other sections to compare the merits and styles of different computational systems. In our own teaching we have experimented with different possibilities, including

- All students in a class using the same system;
- Different students in the same class using different systems;
- Students using *Maple* or *Mathematica* for symbolic computation and MATLAB for numeric computation.

Although each project references the appropriate section of Edwards & Penney, the discussion within each project is largely independent of any particular text. The manual therefore can be used in conjunction with any introductory differential equations textbook that spans the usual range of topics from elementary first-order ordinary differential equations to partial differential equations and boundary value problems.

Each project is designed to provide the basis for an outside class assignment that will engage students for a period of several days (or perhaps a week or two). The various projects may be used in a variety of ways, depending on the specific technology that is available and on the local arrangements for its use. We ourselves have experimented with several formats ranging from homework assignments of moderate length to intensive projects leading to careful written reports.

As both computational technology and its uses in the teaching of differential equations mature, further editions of this manual will be prepared. We will appreciate faculty suggestions as to appropriate revisions and new projects for inclusion, as well as student comments on how our project discussions can be improved for lab use or independent study. These suggestions can be forwarded to us at the following address:

Henry Edwards & David Penney
Department of Mathematics
University of Georgia
Athens, GA 30602-7403

hedwards@math.uga.edu
dpenney@math.uga.edu

Chapter 1

First-Order Differential Equations

Project 1
Slope Fields and Solution Curves

Reference: Section 1.3 of Edwards & Penney
DIFFERENTIAL EQUATIONS with Computing and Modeling

If appropriate computer software is available, it will be instructive for you to generate for yourself the slope fields of Problems 1 through 10 in Section 1.3 of the text. Generate additional slope fields for differential equations $dy/dx = f(x,y)$ of your own choice, and manually sketch some typical solution curves. Assemble a portfolio containing your most interesting and visually attractive results.

A number of specialized differential equations software packages are available, and typically include a facility for the ready construction of slope fields. Systems such as *MacMath* (Macintosh) and *Phaser* (DOS) — as well as the **dfield** function in John Polking's *MATLAB Manual for ODE* (Englewood Cliffs,

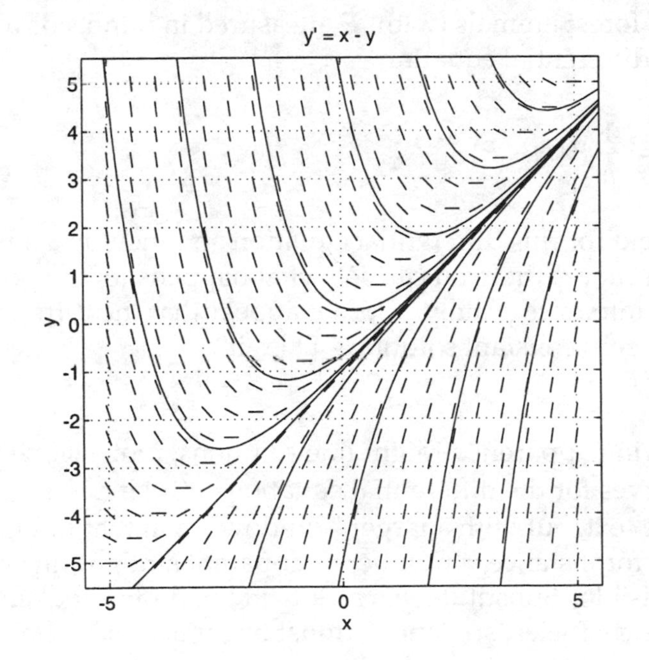

N.J.: Prentice Hall, 1995) — plot both a slope field and, at the click of a mouse button, the solution curve through a desired point.

For example, the figure at the bottom of page 1 shows a slope field and typical solution curves (generated using **dfield**) for the differential equation $y' = x - y$. It appears that there exists a (single) *straight line* solution curve that all other solution curves approach as $x \to \infty$. Indeed, if we substitute the trial straight line solution $y(x) = ax + b$ in the differential equation, we get

$$a = y' = x - y = x - (ax + b) = (1 - a)x - b,$$

which is so if and only if $a = 1$ and $b = -1$. Thus $y(x) = x - 1$ is, indeed, a straight line solution of the differential equation $y' = x - y$. The figure then suggests (without proving) that

$$y(x) - (x - 1) \ \to \ 0 \quad \text{as} \quad x \to \infty.$$

The following two investigations involve similar inferences from slope fields and computer-generated solution curves.

Investigation A
In Section 2.1 of the text we discuss situations in which a population $P(t)$ satisfies the logistic differential equation $P' = aP - bP^2$ (where a and b are constants). For your own personal logistic equation, denote by m and n the *largest* two distinct digits (in either order) in you student ID number. Suppose a population $P(t)$ of forest animals (with P measured in hundreds and t in years) satisfies the differential equation

$$\frac{dP}{dt} = \frac{1}{n} P\,(m - P).$$

Generate a slope field for this differential equation and include a sufficient number of solution curves that you can see what happens to the population as $t \to \infty$. State your inference plainly. Does it appear that the differential equation has a non-zero constant solution? Does it?

Investigation B
For a project involving apparent straight line solutions, plot a slope field and typical solution curves for the differential equation $y' = \sin(x - y)$ of Example 3 in Section 1.3 of the text, but with a larger "window" than that of Figure 1.3.4. With $-8 \le x, y \le 8$, for instance, a number of apparent straight line solution curves should be visible. Substitute $y = ax + b$ in the differential equation to determine what the coefficients a and b must be for a straight line solution

curve. Having done this, can you now infer from your figure how $y(x)$ behaves as $x \to \infty$ (perhaps in terms of the initial value $y(0) = y_0$)?

For your own personal example, let n be the *smallest* integer in your student ID number that is *greater* than 1, and consider the differential equation

$$\frac{dy}{dx} = \frac{1}{n}\cos(x - ny).$$

(i) First investigate the possibility of straight line solutions of this differential equation. That is, determine (in terms of n) all possible values of the constant coefficients a and b such that the linear function

$$y(x) = ax + b$$

is a solution.

(ii) Then generate a slope field for this differential equation, with the viewing window chosen so that you can picture some of these straight lines, plus a sufficient number of non-linear solution curves that you can formulate a conjecture about what happens to $y(x)$ as $x \to \infty$. State your inference as plainly as you can. Given the initial value $y(0) = y_0$, it would be nice to be able to say (perhaps in in terms of y_0) how $y(x)$ appears to behave as $x \to \infty$.

In the sections that follow we outline the use of *Maple*, *Mathematica*, and MATLAB to generate slope fields and solution curves.

Using *Maple*

The differential equation $dy/dx = x - y$ is defined in *Maple* by the command

```
> de1 := diff( y(x), x) = x - y(x);
```

$$de1 := \frac{\partial}{\partial x}y(x) = x - y(x)$$

Note that we write $y(x)$, not just y, to specify that the dependent variable y is a function of the independent variable x, and that **`:=`** is used as the assignment operator. The differential equation solver in *Maple* is **`dsolve`**.

```
> gsoln := dsolve( de1, y(x) );
```

$$gsoln := y(x) = x - 1 + e^{(-x)}\_C1$$

Observe how *Maple* writes the (first) arbitrary constant that appears. We can verify that the righthand side expression in the equation *gsoln* actually satisfies the differential equation by showing that it yields an identity upon substitution.

```
> subs( y(x) = rhs(gsoln), de1 );
```

$$\frac{\partial}{\partial x}(x-1+e^{(-x)}\_C1) = 1-e^{(-x)}\_C1$$

When we enter the command **eval(")** to ask that the derivative on the left be evaluated, the identity

$$1-e^{(-x)}\_C1 = 1-e^{(-x)}\_C1$$

that results shows that the general solution of the differential equation **de1** is indeed $y(x) = x - 1 + Ce^{x}$. We can get a particular solution satisfying a given intitial condition either by specifying it in the **dsolve** command:

```
> psoln1 := dsolve( {de1, y(0)=2}, y(x) );
```

$$psoln1 := \; y(x) = x-1+3e^{(-x)}$$

or by substituting a specific numerical value for the arbitrary constant in the general solution:

```
> psoln2 := subs(_C1=0, gsoln );
```

$$psoln2 := \; y(x) = x-1$$

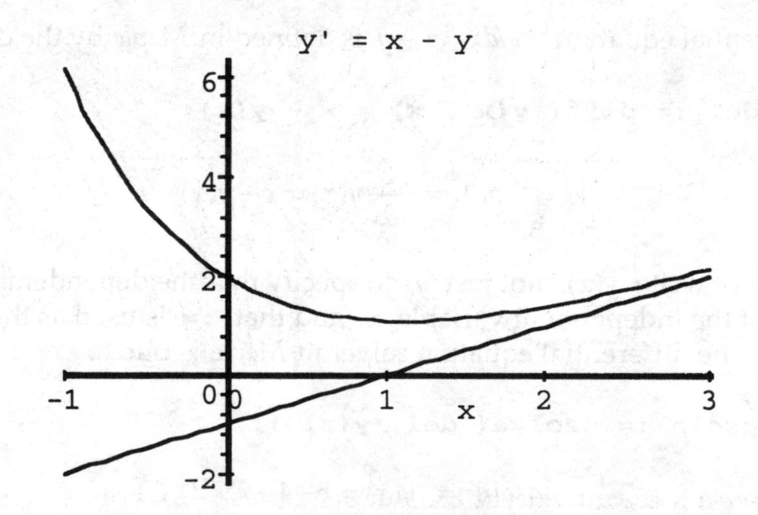

We can plot both these solution curves simultaneously:

```
> y1 := rhs(psoln1):        y2 := rhs(psoln2):

> plot( {y1,y2}, x=-1..3, title = `y' = x - y`);
```

Any finite number of particular solutions could be plotted simultaneously by entering them as a set enclosed by braces. However the **DEtools** package contains a special command for doing this sort of thing.

```
> with(DEtools):
```

DEplot1 is the plotting command for first-order differential equations. The following rather detailed command shows how to plot simultaneously a slope field and a collection of solution curves satisfying a set of separate initial conditions, each corresponding to an initial point (x_0, y_0).

```
> DEplot1( x - y, x=-5..5,
           {[0,4],[0,2],[0,1],[0,0],[0 ,-1],[0,-2],
           [0,-4],[-4,-2],[-3,4],[-4,4],[1,4],[2,4],
           [3,4],[-2,-4],[1,-4],[2,-4],[3,-4]},
           grid=[12,12],
           y=-5..5, title = `y' = x - y`);
```

We have specified the right hand side $f(x, y) = x - y$ of the differential equation $y' = f(x, y)$, the axes $[x, y]$ for plotting, and the window to be plotted.

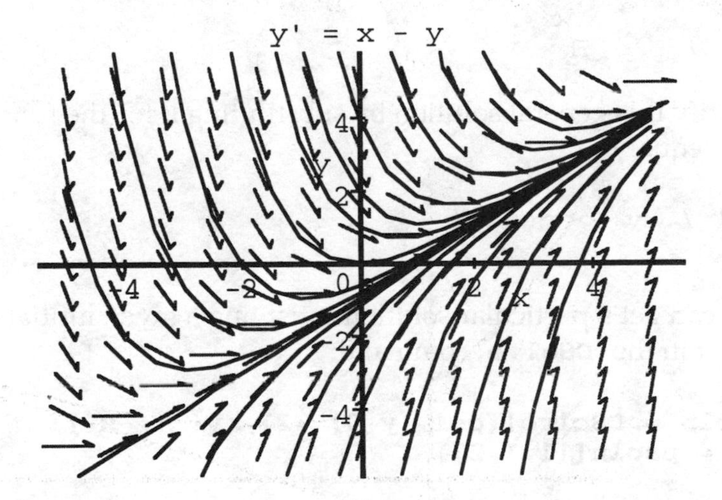

Here we see a pleasant variety of solution curves — all appearing to funnel in on the single linear "asymptotic solution" $y = x + 1$ — together with a slope field consisting of a 12 x12 **grid** of arrows. We could plot the slope field alone by deleting the initial points.

Using *Mathematica*

To define the differential equation $y' = x - y$ in *Mathematica* we enter the command

```
de1 = D[ y[x], x ] == x - y[x]

y'[x] == x - y[x]
```

Note that *Mathematica* uses = for assignment, == for ordinary equality, and square brackets for functional notation. The differential equation solver in *Mathematica* is **DSolve**.

```
desoln = DSolve[ de1, y[x], x ]
                       C[1]
{{y[x] -> -1 + x + ----}}
                        x
                       E
```

Note that *Mathematica* denotes by `C[1]` the (first) arbitrary constant that appears in a general solution, which we may define explicitly as a function of **x** by taking the 2nd element of the 1st element of the 1st element of the nested list **desoln**:

```
gsoln[x_] = desoln[[1,1,2]]
            C[1]
-1 + x + ----
             x
            E
```

We can verify this general solution by substituting it for the function **y** in the differential equation.

```
de1 /. y -> gsoln
True
```

We can get a particular solution satisfying a given intitial condition by specifying it in the **DSolve** command:

```
psoln = DSolve[{de1, y[0]==2}, y[x], x ];
y1 = psoln[[1,1,2]]
      3
-1 + -- + x
      x
     E
```

Or by substituting a numerical value for the arbitrary constant in the general solution:

```
y2 = gsoln[x] /. C[1] -> 0
-1 + x
```

We can plot both these solution curves simultaneously; the command

```
Plot[ {y1,y2}, {x, -1,3} ];
```

produces the same two-curve figure generated previously using *Maple*. Indeed, any finite number of particular solutions could be plotted simultaneously by entering them as a list enclosed by braces. We can construct just such a list by substituting a list (or table) of numerical values for the arbitrary constant `C[1]` in our general solution.

```
yp = gsoln[x] /. C[1] -> Table[c, {c,-5,5} ]
          5              4              3              2
{-1 - -- + x, -1 - -- + x, -1 - -- + x, -1 - -- + x,
          x              x              x              x
         E              E              E              E

       -x                  -x               2
 -1 - E   + x, -1 + x, -1 + E   + x, -1 + -- + x,
                                           x
                                          E

       3              4              5
 -1 + -- + x, -1 + -- + x, -1 + -- + x}
       x              x              x
      E              E              E
```

Here we have specified the values $C[1] = -5,-4, \ldots, 4,5$.

```
curves =
 Plot[ Evaluate[yp], {x,-5,5}, PlotRange -> {-5,5} ];
```

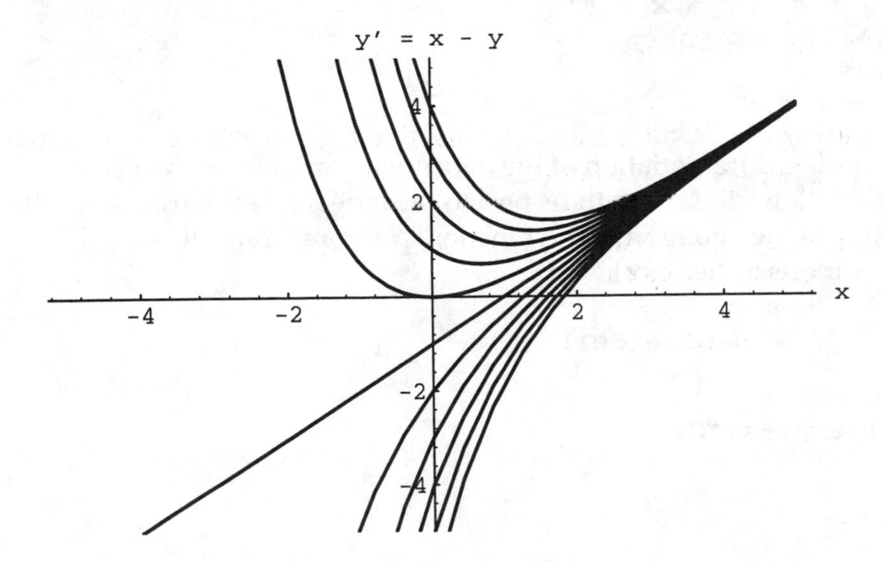

Here we see a family of solution curves, all appearing to funnel in on the single "asymptotic solution" $y = x + 1$. We can use the **PlotVectorField** function, after loading the *Mathematica* package

```
<<Graphics`PlotField`
```

to superimpose a slope field. Then the commands

```
slopes =
PlotVectorField[{1,x-y}, {x,-5,5}, {y,-5,5} ];

Show[ curves, slopes, AspectRatio -> 1 ];
```

do the job; try it for yourself.

Using MATLAB

Figures including slope fields are generated with the greatest ease using the menu-driven **dfield** function in the *MATLAB Manual for ODE* package mentioned previously. Here we illustrate also the "hands on" generation of symbolic solutions and solution curves using the Student Edition of MATLAB (version 4, or the professional edition equipped with the Symbolic Math Toolbox) which calls on an imbedded *Maple* kernel for the execution of symbolic operations.

Symbolic variables, formulas, or equations are entered as strings enclosed in single quotes. Thus we can define the differential equation $y' = x - y$ by entering

```
» de1 = 'Dy = x - y'

de1 =
 Dy = x - y
```

with **D** denoting differentiation of the dependent variable **y** (which immediately follows the **D**) with respect to the independent variable **x** (the other variable in the equation). The function **dsolve** computes explicit symbolic solutions (when possible).

```
» gsoln = dsolve(de1)

gsoln =
 x-1+exp(-x)*C1
```

Apparently we have a general solution involving an arbitrary constant **C1**. To verify this, we check that when we differentiate the expression **y = gsoln** with respect to the symbolic variable **'x'**,

```
» Dy = diff(gsoln,'x')

Dy =
 1-exp(-x)*C1
```

we get the same result as when we substitute **gsoln** for **'y'** in the differential equation,

```
» subs(de1,gsoln,'y')

ans =
 Dy = 1-exp(-x)*C1
```

[The syntax for symbolic substitution of **new** for **old** in the expression **expr** is **subs(expr,new,old)**].

We can get a particular solution satisfying a given intitial condition by specifying it in the **dsolve** command:

```
» psoln1 = dsolve( de1, 'y(0)=2')

psoln1 =
x-1+3*exp(-x)
```

or by substituting a specific numerical value for the arbitrary constant in the general solution:

```
» psoln2 = subs(gsoln, 0, 'C1' )

psoln2 =
x-1
```

We can use MATLAB's **ezplot** command to plot both these two solution curves with the commands

```
» ezplot(psoln1,[-1 3])
» hold on
» ezplot(psoln2,[-1 3])
» axis([-1 3 -2 6])
» grid on
```

The command **ezplot('f', [a b])** plots the expression **'f'** as a function

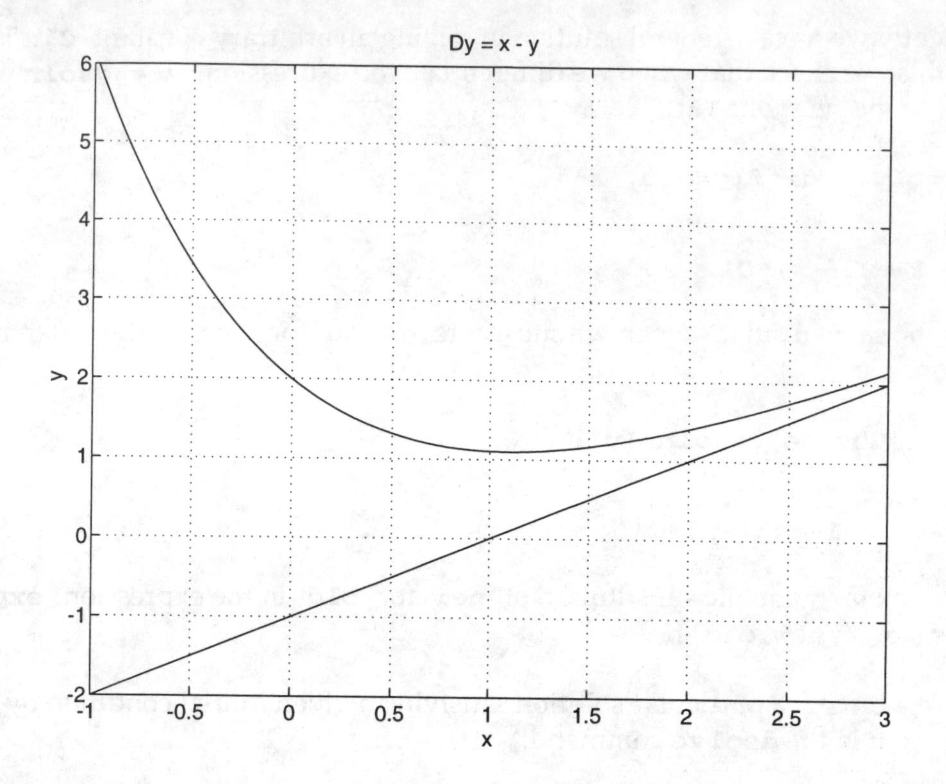

of **x** on the interval $[a, b]$. The **hold on** command holds the first solution curve in place while the while the second one is plotted. The **axis** command overrides MATLAB's autosizing and shows the picture in the desired viewing window:

Unless the special purpose **dfield** program is used, plotting slope fields in MATLAB requiries a bit of work. The following program utilizes the built-in command **quiver(x,y,dx,dy)** that plots an $n \times n$ array of vectors with x- and y- components specified by the $n \times n$ matrices **dx** and **dy**, based at the xy-points in the plane whose x- and y- coordinates are specified by the $n \times n$ matrices **x** and **y**.

```
% slope field program  sfield.m
n = 15;                              % no of subintervals
a = -5; b = 5;                       % x-interval
c = -5; d = 5;                       % y-interval
h = (b-a)/n;   k = (d-c)/n;          % x- and y-step sizes
x = a : h : b;                       % x-subdivision points
y = c : k : d;                       % y-subdivision points
[x,y] = meshgrid(x,y);               % grid of points in plane
f = x - y;                           % define f(x,y)
t = atan(f);                         % angle of inclination
dx = cos(t);     dy = sin(t);        % xy-components of arrow
quiver(x,y,dx,dy)                    % plot slope field
axis([a b c d])                      % viewing window
grid on                              % draw grid lines
```

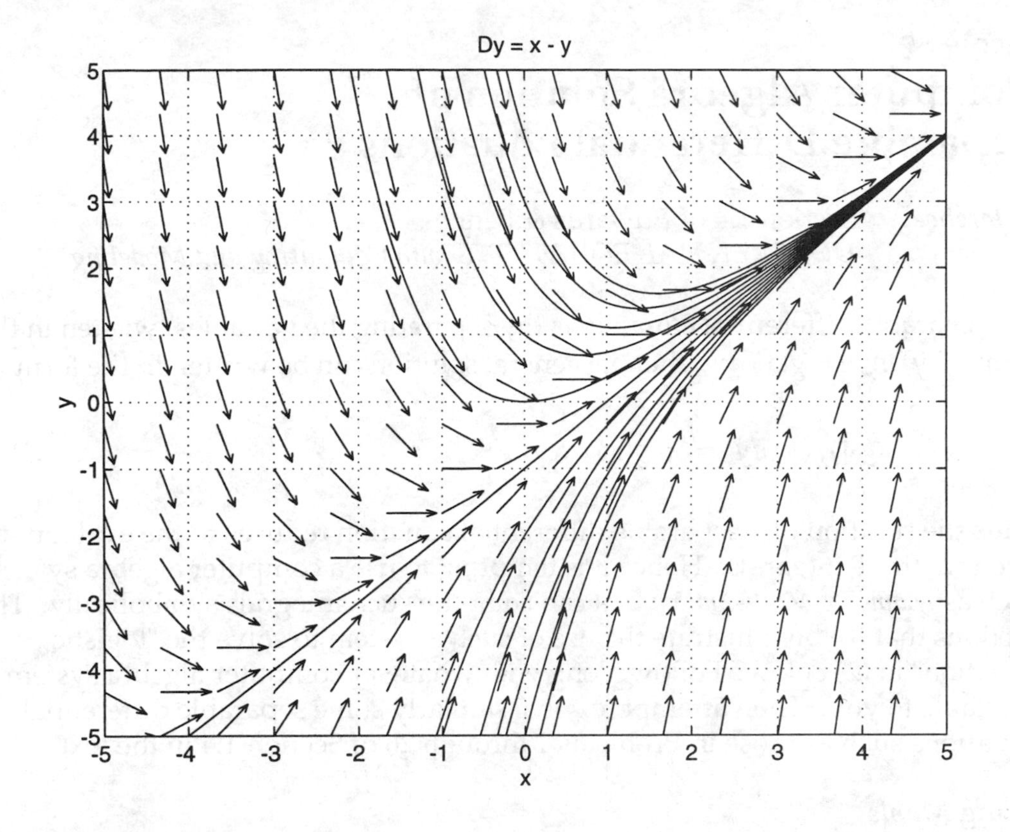

However, you need not be concerned about these matrices of coordinates and components — only the number n of subintervals in each direction, the desired viewing window $a \le b, c \le d,$ and the expression **f** defining the differential equation $y' = f(x,y)$ need be altered when the program **sfield.m** is defined (or when the commands listed above are entered individually in command mode). For instance, once the slope field program **sfield.m** has been defined, the commands

```
» for k = -5 : 5
     ezplot(subs(gsoln,k,'C1'),[-5 5])
     hold on
     end
» sfield
```

first plot the particular solution curves corresponding to the values –5, –4, ..., 4, 5 of the arbitrary constant **C1** in **gsoln**, and then a 15 x 15 slope field of arrows is added.

Project 2

Computer Algebra Solution of Separable Differential Equations

Reference: Section 1.4 of Edwards & Penney
DIFFERENTIAL EQUATIONS with Computing and Modeling

If a separable differential equation is (by separating the variables) written in the form $f(y)\,dy = g(x)\,dx$, then its general solution can be written in the form

$$\int f(y)\,dy = \int g(x)\,dx + C. \qquad (1)$$

Thus the solution of a separable differential equation reduces to the evaluation of two indefinite integrals. Hence it is tempting to use a computer algebra system such as *Maple* or *Mathematica* that can compute such integrals symbolically. The sections that follow illustrate the use of such a system to solve the "logistic equation" in (2) below. Follow along with whatever computer algebra system is available to you. Then attempt to solve similarly some separable differential equations such as those in Problems 1 through 26 of Section 1.4 in the text.

Using *Maple*

For an example of the symbolic algebra solution of a differential equation, consider the logistic equation

$$\frac{dy}{dt} = ky(M-y) \qquad (2)$$

which (as we'll see in Section 2.1 of the text) models the growth of certain human and animal populations. Using *Maple* we separate the variables and write

```
> genSolution := int(1/(y*(M-y)),y) = int(k,t) + C;
```

$$genSolution := \frac{\ln(y)}{M} - \frac{\ln(M-y)}{M} = kt + C$$

The following sequence of successive simplifications is then fairly self-explanatory. First we multiply both sides by M,

```
> genSolution := simplify(M*genSolution);
```

$$genSolution := \ln(y) - \ln(M-y) = M(kt+C)$$

and then we "exponentiate" both sides of the equation.

```
> genSolution := simplify(exp(lhs(")) = exp(rhs(")));
```

$$genSolution := \frac{y}{M-y} = e^{(M(kt+C))}$$

Solution of this equation for y now gives the solution in the explicit form

```
> genSolution := y = solve(genSolution, y);
```

$$genSolution := y = \frac{e^{(M(kt+C))}M}{1+e^{(M(kt+C))}}$$

From the first line of *Maple* output above we see that

```
> C := (ln(y0) - ln(M-y0))/M:
```

in terms of the initial value $y(0) = y_0$. With this value of the arbitrary constant C we get

```
> partSolution := simplify(genSolution);
```

$$partSolution := y = \frac{e^{(ktM)}y0\ M}{M-y0+e^{(ktM)}y0}$$

Thus the particular solution of the logistic equation satisfying the initial condition $y(0) = y_0$ is given by

$$y(t) \;=\; \frac{M\,y0\,e^{ktM}}{M+y0(e^{ktM}-1)} \qquad (3)$$

Using *Mathematica*

We start as in *Maple* by separating the variables and integrating each side of the resulting equation.

```
genSolution =
Integrate[1/(y*(M-y)),y] == Integrate[k,t] + C

Log[y]   Log[-M + y]
------ - ----------- == C + k t
  M          M
```

Note that *Mathematica* is assuming that $y > M$ whereas *Maple* assumed that $y < M$. In *Mathematica* an equation is actually a 2-element list whose first and last elements are the lefthand and righthand sides of the equation. We use this fact to operate separately with the two sides of our equation.

```
genSolution =
Expand[M*First[genSolution]] == M*Last[genSolution]
```

```
Log[y] - Log[-M + y] == M (C + k t)
```

```
genSolution =
Exp[First[genSolution]] == Exp[Last[genSolution]]
```

```
 Log[y] - Log[-M + y]      M (C + k t)
E                      == E
```

```
genSolution =
Expand[First[genSolution]] == Last[genSolution]
```

```
  y         M (C + k t)
------ == E
-M + y
```

Now we can solve for the general solution in explicit form.

```
genSolution = Solve[genSolution, y]
```

```
          C M + k M t
         E            M
{{y -> ------------------}}
                C M + k M t
        -1 + E
```

```
y = First[ y /. genSolution ]
```

```
  C M + k M t
 E            M
-----------------
       C M + k M t
-1 + E
```

To get the particular solution with given initial value $y(0) = y_0$ we substitute the value of C given by the first line of *Mathematica* output above.

```
partSolution = y /. {C -> Log[y0/(y0-M)]/M}
```

```
  k M t + Log[y0/(-M + y0)]
 E                          M
--------------------------------
       k M t + Log[y0/(-M + y0)]
-1 + E
```

The simplifications

```
numer = Simplify[Numerator[partSolution]]

 k M t
E      M y0
-----------
  -M + y0

denom = Simplify[Denominator[partSolution]]

      k M t
     E      y0
-1 + ---------
      -M + y0

partSolution = Simplify[numer/denom]

   k M t
  E      M y0
------------------
          k M t
M - y0 + E      y0
```

finally yield the same formula for the particular solution as given in Equation (3) above.

Using MATLAB

Finally we solve the logistic equation in (2) using the MATLAB interface to the *Maple* kernel. Again we begin by separating variables and integrating each side of the resulting equation. Expressions and variables in MATLAB are denoted by strings delimited with single quotes. The command for integrating an expression with respect to the symbolic variable `'x'` is `int(expression,'x')`. If the variable of integration is omitted, it is by default taken to be the single lower-case letter in the expression, other than `'i'` or `'j'`, that is alphabetically closest to `'x'`.

This time we work separately from the start with the lefthand and righthand sides of the equation. Thus

```
» lhs = int('1/(y*(M-y))')
lhs =
1/M*log(y)-1/M*log(M-y)

» rhs = symadd(int('k','t'),'C')
rhs =
k*t+C
```

On the right we have use the symbolic addition function `symadd` to add the arbitrary constant `'C'`. Next we use the symbolic multiplication function `symmul` to multiply each side by `'M'`.

```
» lhs = simplify(symmul('M',lhs))
lhs =
log(y)-log(M-y)

» rhs = symmul('M',rhs)
rhs =
M*(k*t+C)
```

We now equate the two sides of our equation and solve for the general solution.

```
» equation = [lhs ' = ' rhs]
equation =
log(y)-log(M-y) = M*(k*t+C)

» genSoln = solve(equation,'y')
genSoln =
exp(M*k*t+M*C)*M/(1+exp(M*k*t+M*C))

» pretty
                              exp(M k t + M C) M
                         {y = --------------------}
                              1 + exp(M k t + M C)
```

We see from our equation that $C = (\log(y_0) - \log(M{-}y_0))/M$ if $y(0) = y_0$, so by using the substitution function **subs(expression,new,old)** which substitutes **new** for **old** in the **expression**, we finally get the desired particular solution of the logistic equation.

```
» partSoln = subs(genSoln,'(log(y0)-log(M-y0))/M','C')
partSoln =
exp(M*k*t+log(y0)-log(M-y0))*M/(1+exp(M*k*t+log(y0)-
log(M-y0)))

» partSoln = simplify(partSoln)
ans =
exp(M*k*t)*y0*M/(M-y0+exp(M*k*t)*y0)

» pretty
                                exp(M k t) y0 M
                          ----------------------
                          M - y0 + exp(M k t) y0
```

Project 3

Computer Algebra Solution of Linear First-Order Equations

Reference: Section 1.5 of Edwards & Penney
DIFFERENTIAL EQUATIONS with Computing and Modeling

The text describes the following 4-step algorithm for solving the linear first-order differential equation

$$\frac{dy}{dx} + p(x)\,y = q(x). \tag{1}$$

1. Begin by calculating the *integrating factor*

$$\rho(x) = e^{\int p(x)\,dx}. \tag{2}$$

2. Then multiply both sides of the differential equation by $\rho(x)$.

3. Next, recognize the lefthand side of the resulting equation as the derivative of a product, so it takes the form

$$D_x[\rho(x)\,y(x)] = \rho(x)\,q(x). \tag{3}$$

4. Finally, integrate this last equation to get

$$\rho(x)\,y(x) = \int \rho(x)\,q(x)\,dx + C, \tag{4}$$

and then solve for $y(x)$ to obtain the general solution of the original differential equation in (1).

This algorithim is well-adapted to automatic symbolic computation. We illustrate below its implementation using *Maple*, *Mathematica*, and MATLAB to solve the initial value problem

$$\frac{dy}{dx} - 3\,y = e^{2x}, \qquad y(0) = 3. \tag{5}$$

In each case, the commands shown here constitute a "template" that you can apply to any given linear first-order differential equation. First redefine the coefficient functions $p(x)$ and $q(x)$ as specified by *your* linear equation, then work through the subsequent steps. You may apply whatever computer algebra system is available to carry out this algorithmic process for the initial value problem in (5), and then apply it to a selection of other examples and problems in Section 1.5 of the text.

Using *Maple*

We first define the coefficient functions and initial values specified in (5).

```
> p := -3:
  q := exp(2*x):
  x0 := 0:
  y0 := 3:
```

Then we calculate the integrating factor

```
> rho := exp( int(p,x) );
```

$$\rho := e^{(-3x)}$$

Equation (4) above now gives the general solution

```
> genSolution := y = (1/rho)*(c + int(rho*q, x));
```

$$genSolution := \; y = \frac{c - \dfrac{1}{e^x}}{e^{-(3*x)}}$$

```
> genSolution := simplify(");
```

$$genSolution := \; y = e^{(3*x)}(c - e^{(-x)})$$

If we want a particular solution, we need only substitute the given initial values and solve for the constant c.

```
> c := eval(solve(subs(x=x0,y=y0,genSolution),c));
```

$$c := 4$$

We then evaluate the general solution with this value of the arbitrary constant to get the particular solution

```
> partSolution := eval(genSolution);
```

$$partSolution := \; y = e^{(3*x)}(4 - e^{(-x)})$$

Using *Mathematica*

We first define the coefficient functions and initial values specified in (5).

```
p = -3;
q =  Exp[2x];
x0 = 0;
y0 = 3;
```

Then we calculate the integrating factor

```
rho = Exp[ Integrate[p,x] ]

 -3 x
E
```

Equation (4) above now gives the general solution

```
genSolution = y == (1/r)(c + Integrate[rho*q,x])

        3 x         -x
y == E      (c - E   )
```

If we want a particular solution, we need only substitute the given initial values and solve for the constant *c*.

```
Solve[genSolution, c] /. {x -> x0, y -> y0}

{{c -> 4}}
```

When we substitute this value of the arbitrary constant in the general solution, we get the desired particular solution

```
partSolution = genSolution /. c -> First[c /. %]

        3 x         -x
y == E      (4 - E   )
```

Using MATLAB

First we define the coefficient functions and initial values

```
» p = '-3';
» q = 'exp(2*x)';

»x0 = '0';        y0 = '3';
```

that appear in (5), using symbolic rather than numeric constants. To build up the general solution in (4), we use the symbolic operations **symadd(a,b)** and **symmul(a,b)** for adding and multiplying the symbolic expressions **a** and **b**.

```
» recip_rho = symdiv('1',rho)
recip_rho =
1/(E^(-3*x))

» integral = int(symmul(rho,q))
integral =
-1/exp(x)

» genSolution = symmul(recip_rho,symadd('c',integral))
genSolution =
1/(E^(-3*x))*(c-1/exp(x))

» genSolution = simplify(genSolution)
genSolution =
exp(3*x)*(c-exp(-x))
```

Thus we have the general solution

$$y(x) = e^{3x}\left(c - e^{-x}\right)$$

To find the desired particular solution, we set up the equation to solve for the numerical value of the arbitrary constant c.

```
» equation = [y0,' = ',subs(genSolution,x0,'x')]
equation =
3 = exp(0)*(c-exp(0))
```

and then solve it:

```
» c0 = solve(equation,'c')
c0 =
4
```

Thus our particular solution is given by

```
» partSolution = subs(genSolution,c0,'c')
partSolution =
exp(3*x)*(4-exp(-x))
```

That is,

$$y(x) = e^{3x}\left(4 - e^{-x}\right) = 4e^{3x} - e^{2x}$$

Project 4

Computer Algebra Implementation of Substitution Techniques

Reference: Section 1.6 of Edwards & Penney
DiIFFERENTIAL EQUATIONS with Computing and Modeling

As illustrated below, the substitutions of Section 1.6 in the text are readily carried out using a computer algebra system such as *Maple, Mathematica*, or MATLAB. Thus you can use whatever computer algebra system is available to lighten the computational labor of carrying out the substitutions required to solve Problems 1-30 in the text.

Using *Maple*

The differential equation

$$\frac{dy}{dx} = (x+y+3)^2 \tag{1}$$

of Example 1 in the text calls for the substitution

$$v = x+y+3 \tag{2}$$

We enter first our differential equation (1) in terms of y,

```
> y_diffeq := diff(y(x),x) = (x + y(x) + 3)^2;
```

$$y\_diffeq := \frac{\partial}{\partial x} y(x) = (x+y(x)+3)^2$$

and then the substitution

```
> v_subst := v(x) = x + y(x) + 3;
```

$$v\_subst := v(x) = x + y(x) + 3$$

in (2) for v in terms of y. The inverse substitution for y in terms of v will be

```
> subst_y := y(x) = solve(subst_v, y(x));
```

$$subst\_y := y(x) = v(x) - x - 3$$

After calculating the derivative

```
> Dy_subst := diff(subst_y,x);
```

$$Dy\_subst := \frac{\partial}{\partial x}y(x) = \left(\frac{\partial}{\partial x}v(x)\right) - 1$$

we substitute for $y(x)$ and $y'(x)$ in the original y-equation to get our transformed differential equation

```
> v_diffeq := eval(subs(y_subst, Dy_subst, y_diffeq));
```

$$v\_diffeq := \left(\frac{\partial}{\partial x}v(x)\right) - 1 = v(x)^2$$

to be solved for $v(x)$. Having gotten this far with the computer, we might as well call on *Maple*'s **dsolve** function to solve this differential equation.

```
> v_solution := dsolve( v_diffeq, v(x) );
```

$$v\_solution := -\arctan(v(x)) + x = \_C1$$

It remains only to re-substitute for v in terms and y and solve explicitly for the solution $y(x)$ of our original differential equation.

```
> y_solution := subs(v_subst, v_solution );
```

$$y\_solution := -\arctan(x + y(x) + 3) + x = \_C1$$

```
> y(x) = solve( y_solution, y(x) );
```

$$y(x) = \tan(x - \_C1) - x - 3$$

The *Maple* commands shown above provide a "template" that can be used with more complicated differential equations. You need only re-enter your own differential equation **y_diffeq** and your desired substitution **v_subst**, and then proceed to re-execute the remaining commands.

Using *Mathematica*

Here we illustrate the use of *Mathematica* to solve a homogeneous differential equation of the form

$$\frac{dy}{dx} = F\left(\frac{y}{x}\right) \qquad (3)$$

using the standard substitutions

$$v = \frac{y}{x}, \qquad y = vx, \qquad \frac{dy}{dx} = v + x\frac{dv}{dx}. \tag{4}$$

To solve the differential equation

$$2xy\frac{dy}{dx} = 4x^2 + 3y^2 \tag{5}$$

of Example 2 in the text, for instance, we first enter the equation as

```
Clear[y,v]
yDiffeq  =  2 x y[x] y'[x] == 4 x^2 + 3 y[x]^2

                       2           2
2 x y[x] y'[x] == 4 x   + 3 y[x]
```

The substitution

```
y[x_] := x v[x]
```

in (4) then yields the transformed differential equation

```
vDiffeq = yDiffeq

   2                                   2        2       2
2 x  v[x] (v[x] + x v'[x]) == 4 x  + 3 x  v[x]
```

when we re-evaluate **yDiffeq**. Since we're prodeeding entirely by computer, let's just use *Mathematica*'s **DSolve** function to solve for $v(x)$.

```
vSolution = DSolve[vDiffeq, v[x] ,x ]

{{v[x] -> -Sqrt[-4 + x C[1]]},
 {v[x] ->  Sqrt[-4 + x C[1]]}}

v[x_] = v[x] /. vSolution

{-Sqrt[-4 + x C[1]], Sqrt[-4 + x C[1]]}
```

We get two distinct v-solutions differing by sign, and the corresponding y-solutions result when we re-evaluate $y(x)$, thereby automatically substituting each formula for $v(x)$ via $y(x) = x\,v(x)$.

```
ySolution = y[x]

{-(x Sqrt[-4 + x C[1]]), x Sqrt[-4 + x C[1]]}
```

The implicit solution $y^2 + 4x^2 = Cx^3$ found in the text results when we square both sides.

```
implicitSolution =
y^2 == Expand[Simplify[ First[y[x]]^2 ]] /. C[1] -> C

 2          2        3
y   == -4 x   + C x
```

The *Mathematica* commands shown above provide a "template" that can be applied to any other homogeneous differential equation. You need only re-enter your own homogeneous differential equation **yDiffeq** and then proceed to re-execute the remaining commands.

Using MATLAB

The differential equation

$$\frac{dy}{dx} = (x+y+3)^2 \tag{1}$$

of Example 1 in the text calls for the substitution

$$v = x+y+3 \tag{2}$$

We enter first our differential equation (1) in terms of y,

```
» y_diffeq = 'Dy = (x + y + 3)^2'
y_diffeq =
Dy = (x + y + 3)^2
```

and the substitution

```
» v_subst = 'x + y + 3'
v_subst =
x + y + 3
```

for v in terms of y. To invert this substitution, we set up the equation

```
» v_subst_eq = ['v = ',v_subst]
v_subst_eq =
v = x + y + 3
```

and solve for y in terms of v:

```
» y_subst = solve(subst_eq, 'y')
y_subst =
v-x-3
```

We calculate the derivative

```
» Dy_subst = 'Dv - 1'
Dy_subst =
Dv - 1
```

of y in terms of the derivative of v, and then transform the original differential equation **y_diffeq** by successively substituting **y_subst** for **y** and **Dy_subst** for **Dy**.

```
» temp_diffeq = subs(y_diffeq,y_subst,'y')
temp_diffeq =
Dy = v^2

» v_diffeq = subs(temp_diffeq,Dy_subst,'Dy')
v_diffeq =
Dv-1 = v^2
```

Now MATLAB's **dsolve** function gives

```
» v_solution = dsolve(v_diffeq)
v_solution =
-tan(-x+C1)
```

Finally, we can solve the equation

```
» v_solution_eq = [v_subst,' = ',v_solution]
v_solution_eq =
x + y + 3 = -tan(-x+C1)
```

for the solution

```
» y_solution = solve(v_solution_eq,'y')
y_solution =
-x-3+tan(x-C1)
```

of our original differential equation. This solution is given explicitly by

```
» y_solution_eq = ['y(x) = ',y_solution]
y_solution_eq =
y(x) = -x-3+tan(x-C1)
```

The MATLAB commands shown above provide a "template" that can be used with more complicated differential equations. You need only re-enter your own differential equation **y_diffeq** and your desired substitution **v_subst**, and then proceed to re-execute the remaining commands.

Project 5

Classification of First-Order Differential Equations

Reference: Sections 1.4–1.6 of Edwards & Penney
DiIFFERENTIAL EQUATIONS with Computing and Modeling

In preceding projects we have mentioned the "dsolve" commands in *Maple, Mathematica*, and MATLAB for the symbolic solution of first-order differential equations. These are "black-box" commands that can be applied without inquiring as to the specific type of the differential equation whose solution is sought.

The widely available computer algebra system *Derive* has some features that are especially useful for review purposes. The *Derive* utility file ODE1.MTH contains separate commands for the solution of separable, linear, exact, homogeneous, and Bernoulli equations (among others). Thus the user must first classify the particular differential equation at hand, then apply the appropriate *Derive* command. The pertinent commands are shown in the following table, together with the form and type of equation to which each applies

Form of Equation	*Type*	*Derive Command*
$y' = p(x)q(y)$	Separable	**SEPARABLE_GEN(p(x),q(y),x,y)**
$y' + p(x)y = q(x)$	Linear	**LINEAR_GEN(p(x),q(x),x,y)**
$y' + p(x)y = q(x)y^k$	Bernoulli	**BERNOULLI_GEN(p(x),q(x),k,x,y)**
$y' = r(x,y) = F(y/x)$	Homogen.	**HOMOGENEOUS_GEN(r(x,y),x,y)**
$p(x,y)dx + q(x,y)dy = 0$	Exact	**EXACT_GEN(p(x,y),q(x,y),x,y)**

The suffix **GEN** in each case indicates that a general solution (perhaps implicit) containing an arbitrary constant **c** is produced. If this suffix is omitted and initial values are inserted then a particular solution results. For instance, the command

```
SEPARABLE(p(x),q(y),x,y,x0,y0)
```

generates the particular solution satisfying the initial condition $y(x_0) = y_0$.

Example 1 The differential equation $(y^2+1)dx + 2xy\,dy = 0$ is simultaneously exact, separable, and a Bernoulli equation. The *Derive* command

```
EXACT_GEN(y^2 + 1, 2 x y, x, y)
```

yields after simplification the solution $x(y^2 + 1) = c$, which results also (in equivalent forms) from the commands

```
SEPARABLE_GEN(-1/(2x), (y^2 + 1)/y, x, y)
```

and

```
BERNOULLI_GEN(1/(2x), -1/(2x), -1, x, y)
```

Do you see where the entries in these last two commands came from?

If *Derive* is available, solve each of the following differential equations in both of the indicated ways.

1. $y' = 3(y+7)x^2$ (separable and linear)

2. $y' + xy = xy^3$ (separable and Bernoulli)

3. $(2y^2 + 3x^2)dx + 4xy\,dy = 0$ (exact and homogeneous)

4. $(x+3y)dx + (3x-y)\,dy = 0$ (exact and homogeneous)

5. $(x^2+1)y' = 2x(y+1)$ (separable and linear)

6. $(\tan x)y' = \sqrt{y} - y$ (separable and Bernoulli)

Even if *Derive* software is not available, you still can analyze the equations and write the two pertinent *Derive* commands. Another review approach is to solve these equations using the computer algebra methods of Projects 2–4. *Maple*, *Mathematica*, and MATLAB techniques for the symbolic solution of exact first-order equations are illustrated below.

It may be even more instructive for you to produce your own examples of such differential equations – ones that are simultaneously separable and linear, exact and Bernoulli, etc. See how many such combinations you can achieve.

Exact Equations

Recall that, in order to solve the equation

$$M(x,y)\,dx + N(x,y)\,dy = 0 \tag{1}$$

with $\partial M / \partial y = \partial N / \partial x$, we integrate $M(x, y)$ with respect to x and write

$$F(x,y) = \int M(x,y)\,dx + g(y), \tag{2}$$

thinking of the function $g(y)$ as an "arbitrary constant of integration" as far as the variable x is concerned. Then we determine $g(y)$ by imposing the condition that

$$\frac{\partial F}{\partial y} = N(x,y). \tag{3}$$

This yields a general solution in the implicit form $F(x, y) = C$.

Using *Maple*

In order to solve the differential equation

$$(6xy - y^3)dx + (4y + 3x^2 - 3xy^2)dy = 0 \tag{4}$$

or Example 9 in Section 1.6 of the text, we first define the coefficient functions in (1),

```
> M := 6*x*y - y^3:    N := 4*y + 3*x^2 - 3*x*y^2:
```

and then check the exactness criterion:

```
> diff(M,y) = diff(N,x);
```

$$6x - 3y^2 = 6x - 3y^2$$

We can therefore proceed with the integral in (2),

```
> F := int(M, x) + g(y);
```

$$F := 3x^2y - y^3x + g(y)$$

Then Eq. (3) takes the form

```
> eq3 := diff(F,y) = N;
```

$$eq3 := 3x^2 - 3xy^2 + \left(\frac{\partial}{\partial y}g(y)\right) = 4y + 3x^2 - 3xy^2$$

We first solve this equation for the derivative

```
> Dg := solve(eq3, diff(g(y),y));
```

$$Dg := 4y$$

and then integrate with respect to y to obtain

```
> g(y) := int(Dg,y);
```

$$g(y) := 2y^2$$

Finally, the implicit solution $F(x, y) = C$ takes the form

```
> F = C;
```

$$3x^2y - y^3x + 2y^2 = C$$

Using *Mathematica*

In order to solve the differential equation in (4), we first define the coefficient functions in (1),

```
m := 6x y - y^3;     n := 4y + 3x^2 - 3x y^2
```

(using lower-case **m** and **n** because the symbol **N** is reserved in *Mathematica*) and then check the exactness criterion

```
D[m, y] == D[n, x]

True
```

We can therefore proceed with the integral in (2),

```
F = Integrate[m, x] + g[y]

   2        3
3 x  y - x y  + g[y]
```

Then Eq. (3) takes the form

```
eq3 =   D[F, y] == n

   2          2                   2                 2
3 x  - 3 x y  + g'[y] == 3 x  + 4 y - 3 x y
```

We first solve this equation for the derivative

```
solution = Solve[eq3, g'[y]]

{{g'[y] -> 4 y}}
```

```
Dg = First[g'[y] /. solution]
4 y
```

and then integrate with respect to y to obtain

```
g[y] = Integrate[Dg, y]
   2
2 y
```

Finally, the implicit solution $F(x, y) = C$ takes the form

```
F == C

   2          2        3
3 x  y + 2 y  - x y   == C
```

Using MATLAB

In order to solve the differential equation in (4), we first define the coefficient functions in (1),

```
» M = '6*x*y-y^3';        N = '4*y+3*x^2-3*x*y^2';
```

and then check the exactness criterion:

```
» diff(M,'y')
ans =
6*x-3*y^2

» diff(N,'x')
ans =
6*x-3*y^2
```

We can therefore proceed with the integral in (2),

```
» F = int(M,'x')
F =
3*x^2*y-y^3*x
```

Here `F` denotes the integral in (2) *without* the arbitrary function $g(y)$, so Eq. (3) now takes the form

$$\frac{\partial F}{\partial y} + g'(y) = N(x,y),$$

so $g'(y) = N(x,y) - \partial F / \partial y$:

```
» DF = diff(F,'y')
DF =
3*x^2-3*x*y^2

» eq3 = symsub(N,DF)
eq3 =
4*y
```

Thus $g'(y) = 4y$ so the function $g(y)$ is given by

```
» g = int(eq3,'y')
g =
2*y^2
```

Finally, when we set up the equation

$$F(x,y)+g(y) = C$$

we get the implicit solution

```
» solution = [F,' + ',g,' = C']
solution =
3*x^2*y-y^3*x + 2*y^2 = C

» pretty(solution)
                                      2      3         2
                                   3 x  y - y  x + 2 y   = C
```

of the exact differential equation

$$(6xy-y^3)dx+(4y+3x^2-3xy^2)dy = 0$$

in (4).

Chapter 2

Mathematical Models and Numerical Methods

Project 6
Fitting Data to Logistic Population Models

Reference: Section 2.1 of Edwards & Penney
DIFFERENTIAL EQUATIONS with Computing and Modeling

This project deals with the problem of fitting a logistic model to given population data. Thus we want to determine the numerical constants a and b so that the solution $P(t)$ of the initial value problem

$$\frac{dP}{dt} = aP + bP^2, \qquad P(0) = P_0 \tag{1}$$

approximates the given values $P_0, P_1, \ldots, P_n$ of the population at the times $t_0 = 0, t_1, \ldots, t_n$. If we rewrite Eq. (1) — the logistic equation with $kM = a$ and $k = -b$ — in the form

$$\frac{1}{P}\frac{dP}{dt} = a + bP, \tag{2}$$

then we see that the points

$$\left(P(t_i), \frac{P'(t_i)}{P(t_i)} \right) \qquad i = 0, 1, 2, \ldots, n,$$

should all lie on the straight line with y-intercept a and slope b that is defined by the linear function of P that appears on the right-hand side in Eq. (2).

This observation provides a way to find the values of a and b. If we can determine the approximate values of the derivatives $P_1', P_2', P_3', \ldots$ corresponding to the given population data, then we can proceed with the following agenda.

- First plot the points $(P_1, P_1'/P_1)$, $(P_2, P_2'/P_2)$, $(P_3, P_3'/P_3)$, $\ldots$ on a sheet of graph paper with horizontal P-axis.

- Then use a ruler to draw a straight line that appears to approximate these points well.

- Finally measure this straight line's y-intercept a and slope b.

But where are we to find the needed values of the derivative $P'(t)$ of the (as yet) unknown function $P(t)$? It is easiest to use the approximation

$$P_i' = \frac{P_{i+1} - P_{i-1}}{t_{i+1} - t_{i-1}} \qquad (3)$$

that is suggested by Fig. 2.1.6 in the text. Note that, if the t-values for the given data points are equally distributed with successive difference $h = t_{i+1} - t_{i-1}$, then the quotient on the right-hand side in (3) is a difference quotient of the type

$$\frac{\Delta P}{\Delta t} = \frac{P(t_i + h) - P(t_i - h)}{2h}$$

that is used to define the derivative $P'(t)$, and is "centered" at the point t_i. For instance, using the U.S. population data given in Fig. 2.1.2 in the text, we obtain these results:

For 1800: $$P_1' = \frac{7.2 - 3.9}{20} = 0.165$$

For 1810: $$P_2' = \frac{9.6 - 5.3}{20} = 0.215$$

$$\vdots$$

For 1920: $$P_{13}' = \frac{123.2 - 92.2}{20} = 1.550$$

Investigation A

Finish the computation indicated above, then plot on a sheet of graph paper the points $(P_1, P_1'/P_1)$, $(P_2, P_2'/P_2)$, $\ldots, (P_{13}, P_{13}'/P_{13})$ as indicated by the small circles in Fig. 2.1.7 in the text. Draw your own straight line approximating these points, then measure its y-intercept and slope as accurately as you can. Solve Eq. (1) with the resulting numerical parameters a and b. How do your predicted populations for the years 1800 through 2000 compare with the population data given in Figs. 2.1.2 and 2.1.8?

Investigation B

Repeat Investigation A, but take $t = 0$ in 1900 and use only 20th century population data. Do you get a better approximation for the U.S. population during the final decades of the 20th century?

Investigation C

Model similarly the world population data shown in Fig. 2.1.9 in the text. The Population Division of the United Nations predicts a world population of 8.177 billion in the year 2025. What do you predict?

The Method of Least Squares

So you think there ought to be a better way than eyeballing the placement of a ruler on a sheet a paper? There is, and it dates back to Gauss. Writing Q for the left-hand side in Eq. (2), we seek the straight line

$$Q = a + bP \tag{4}$$

in the PQ-plane that "best fits" n given data points (P_1, Q_1), (P_2, Q_2), ..., (P_n, Q_n). Gauss' idea was to choose this line so that it minimizes the sum of the squares of the vertical distances between the line and these points. The vertical distance — the difference in Q-coordinates above P_i — between the ith point (P_i, Q_i) is $d_i = Q_i - (a + bP_i)$, so the sum of the squares of these "errors" is the function

$$f(a,b) = \sum_{i=1}^{n} [Q_i - (a + bP_i)]^2 \tag{5}$$

of the (as yet) unknown parameters a and b.

To minimize the value $f(a,b)$ as a function of a and b, we calculate the partial derivatives

$$\frac{\partial f}{\partial a} = \sum_{i=1}^{n} 2[Q_i - (a + bP_i)](-1) = 2a\sum_{i=1}^{n} 1 + 2b\sum_{i=1}^{n} P_i - 2\sum_{i=1}^{n} Q_i$$

and

$$\frac{\partial f}{\partial b} = \sum_{i=1}^{n} 2[Q_i - (a + bP_i)](-P_i) = 2a\sum_{i=1}^{n} P_i + 2b\sum_{i=1}^{n} P_i^2 - 2\sum_{i=1}^{n} P_i Q_i.$$

When we set both partial derivatives equal to zero, we get the pair

$$a\,n+b\sum_{i=1}^{n}P_i \;=\; \sum_{i=1}^{n}Q_i \tag{6}$$

$$a\sum_{i=1}^{n}P_i+b\sum_{i=1}^{n}P_i^2 \;=\; \sum_{i=1}^{n}P_iQ_i \tag{7}$$

of *linear* equations in the unknowns a and b, with coefficients that are simple combinations of the *P*- and *Q*-coordinates of the n given data points. It remains only to solve Eqs. (6)–(7) for the coefficients a and b in the desired straight line (4) that best fits these data points.

In the following paragraphs we illustrate this least squares approach using *Maple*, *Mathematica*, and MATLAB. In each we consider the $n = 6$ given data points

$$(1.1, 2.88),\ (1.9, 4.36),\ (3.2, 6.98),\ (4.6, 9.86),\ (5.3, 11.20),\ (6.5, 13.54).$$

You can automate similarily your calculations for Investigations A, B, and C.

Using *Maple*

To set up Eqs. (6) and (7) we first enter the lists

```
> P := [1.1, 1.9, 3.2, 4.6, 5.3, 6.5]:
> Q := [2.88, 4.36, 6.98, 9.86, 11.20, 13.54]:
```

of the *P*- and *Q*-coordinates of the

```
> n := 6:
```

given data points. Next we use the *Maple* **sum** function to calculate the sums that appear as coefficients of a and b in Eqs. (6)-(7).

```
> sumP := evalf(sum( P[i], i = 1..6)):
> sumQ := evalf(sum( Q[i], i = 1..6)):
> sumPsq := evalf(sum( P[i]^2, i = 1..6)):
> sumPQ := evalf(sum( P[i]*Q[i], i = 1..6)):
```

Then the two equations we want to solve are defined by

```
> eq6 := n*a + sumP*b = sumQ;
```

$$eq6 := \ 6\,a + 22.6\,b \ = \ 48.82$$

```
> eq7 := sumP*a + sumPsq*b = sumPQ;
```

$$eq7 := \ 22.6\,a + 106.56\,b \ = \ 226.514$$

Now we simply solve these two equations for the coefficients a and b using the command

```
> soln := fsolve({eq6,eq7}, {a,b});
```

$$\{a = .6457449574,\ b = 1.988740277\}$$

```
> a := rhs(soln[1]):        b := rhs(soln[2]):
```

Thus the least squares best fit in Eq. (4) takes the form

$$Q \ = \ 0.6457 + \ 1.9887\,P$$

(rounding the coefficients to four decimal places). Finally the differences between the actual values $\{Q_i\}$ and the predicted values $\{0.6457 + \ 1.9887\,P_i\}$ are given by

```
> seq(Q[i] - (a + b*P[i]), i = 1..6);
```

$$.0466, -.0644, -.0297, .0660, .0139, -.0326$$

(again rounding to four decimal places).

Using *Mathematica*

To set up Eqs. (6) and (7) we first enter the lists

```
P = {1.1, 1.9, 3.2, 4.6, 5.3, 6.5};

Q = {2.88, 4.36, 6.98, 9.86, 11.20, 13.54};
```

of the P- and Q-coordinates of the

```
n = 6;
```

given data points. Next we use the *Mathematica* `Sum` function to calculate the sums that appear as coefficients of a and b in Eqs. (6)-(7).

```
sumP = Sum[ P[[i]], {i, 1,6} ];
sumQ = Sum[ Q[[i]], {i, 1,6} ];
sumPsq = Sum[ P[[i]]^2, {i, 1,6} ];
sumPQ = Sum[ P[[i]]*Q[[i]], {i, 1,6} ];
```

Then the two equations we want to solve are defined by

```
eq6 = n*a + sumP*b == sumQ
6 a + 22.6 b == 48.82
eq7 = sumP*a + sumPsq*b == sumPQ
22.6 a + 106.56 b == 226.514
```

Now we simply solve these two equations for the coefficients a and b using the command

```
soln = NSolve[{eq6,eq7}, {a,b}]
{{a -> 0.645745, b -> 1.98874}}
a = First[a /. soln];  b = First[b /. soln];
```

Thus the least squares best fit in Eq. (4) takes the form

$$Q = 0.6457 + 1.9887\,P$$

(rounding the coefficients to four decimal places). Finally the differences between the actual values $\{Q_i\}$ and the predicted values $\{0.6457 + 1.9887\,P_i\}$ are given by

```
Q - (a + b P)
{0.0466, -0.0644, -0.0297, 0.0660, 0.0139, -0.0326}
```

(again rounding to four decimal places).

Using MATLAB

To set up Eqs. (6) and (7) we first enter the lists

```
» P = [1.1, 1.9, 3.2, 4.6, 5.3, 6.5];
» Q = [2.88, 4.36, 6.98, 9.86, 11.20, 13.54];
```

of the P- and Q-coordinates of the

```
» n = 6;
```

given data points. Next we use the MATLAB **sum** function to calculate the sums that appear as coefficients of a and b in Eqs. (6)-(7).

```
» sumP = sum(P);

» sumQ = sum(Q);

» sumPsq = sum(P.*P);

» sumPQ = sum(P.*Q);
```

Then the two symbolic equations we want to solve are defined by

```
» eq6 = [symop(n,'*','a','+',sumP,'*','b'),
                                   ' = ',symmul('1',sumQ)]
eq6 =
6*a+113/5*b = 2441/50

» eq7 = [symop(sumP,'*','a','+',sumPsq,'*','b'),
                                   ' = ',symmul('1',sumPQ)
eq7 =
113/5*a+2664/25*b = 1992438214832423/8796093022208
```

Now we simply solve these two equations for the coefficients a and b using the command

```
» [a,b] = solve(eq6,eq7,'a,b');

» a = numeric(a)
a =
     0.6457

» b = numeric(b)
b =
     1.9887
```

Thus the least squares best fit in Eq. (4) takes the form

$$Q \;=\; 0.6457 + \; 1.9887\, P$$

(with the coefficients rounded to four decimal places). Finally the differences between the actual values $\{Q_i\}$ and the predicted values $\{0.6457 + \; 1.9887\, P_i\}$ are given by

```
» Q - (a + b*P)
ans =
     0.0466   -0.0644   -0.0297    0.0660    0.0139   -0.0326
```

(again rounded to four decimal places).

Project 7

Computer Algebra Solution of Autonomous Equations by Integration

Reference: Section 2.2 of Edwards & Penney *DIFFERENTIAL EQUATIONS with Computing and Modeling*

Every autonomous first-order differential equation

$$\frac{dx}{dt} = f(x) \tag{1}$$

is immediately separable, so its solution reduces to an integration:

$$t = \int \frac{dx}{f(x)}. \tag{2}$$

For instance, consider the case

$$\frac{dx}{dt} = (x-a)(x-b)(x-c) \tag{3}$$

in which $f(x)$ is a cubic polynomial with distinct real zeros a, b, and c. Then the integral

$$t = G(x) = \int \frac{dx}{(x-a)(x-b)(x-c)} \tag{4}$$

can be evaluated explicitly using a computer algebra system such as *Maple*, *Mathematica*, or MATLAB — or even manually using a partial fractions decomposition of the form

$$\frac{1}{(x-a)(x-b)(x-c)} = \frac{A}{x-a} + \frac{B}{x-b} + \frac{C}{x-c}. \tag{5}$$

Thus the resulting antiderivative will have the form

$$G(x) = A\ln|x-a| + B\ln|x-b| + C\ln|x-c|. \tag{6}$$

The corresponding formal solution of Eq. (3) is then given by

$$G(x) = t + K \tag{7}$$

where K is a constant of integration.

The particular solution $x = x(t)$ of Eq. (3) satisfying the initial condition $x(t_0) = x_0$ is defined implicitly by the equation

$$G(x) = t + G(x_0) - t_0. \tag{8}$$

But instead of attempting to solve for x as a function, why not take the easy way out and solve for t as a function of x:

$$t = t_0 + G(x) - G(x_0). \tag{9}$$

If a computer graphing utility is available, then "solution curves" of Eq. (3) are readily plotted in the xt-plane with horizontal x-axis and vertical t-axis. Assuming that this has been done, is it clear how to convert the resulting figure into a picture showing "ordinary" solution curves in the tx-plane with horizontal t-axis and vertical x-axis?

For your own individual project, carry out the program described above with $|a|$, $|b|$, and $|c|$ being the three smallest distinct nonzero digits in your student I.D. number; choose signs for a, b, and c however you wish. If you're more ambitious numerically, you can solve in this manner a differential equation such as

$$\frac{dx}{dt} = x^3 - 3x^2 + 1 \tag{10}$$

whose right-hand side cubic polynomial has distinct roots that are real but irrational. In this case you'll have to begin by solving numerically for the distinct real solutions a, b, and c of the equation

$$f(x) = x^3 - 3x^2 + 1 = 0. \tag{11}$$

Figure 2.2.6 in the text shows four curves of the form $t = f(x)$ plotted on each of the intervals $(-3, a)$, (a, b), (b, c), and $(c, 5)$ with $a < b < c$, together with the vertical asymptotes $x = a$, $x = b$, and $x = c$. When you convert this xt-figure into a tx-figure — simply by reflecting in the line $x = t$? — do you see two spouts and a funnel?

Using *Maple*

With a computer algebra system, the integral in (4) is readily at hand if the zeros a, b, c of the cubic denominator are known. For instance:

```
> int(1/((x+3)*(x-1)*(x-2)), x);
```

$$\frac{1}{20}\ln(x+3)-\frac{1}{4}\ln(x-1)+\frac{1}{5}(x-2)$$

Note that *Maple* does not mention the absolute values, so we must supply the correct signs ourselves, together with the arbitrary constant. For instance, a solution of the initial value problem

$$x' = (x+3)(x-1)(x-2), \qquad x(0) = 0$$

in the open interval (–3, 1) is defined implicitly by

```
> t := x -> (1/20)*ln(x+3)-(1/4)*ln(1-x)+(1/5)*ln(2-x)-
            (1/20)*ln(3)-(1/5)*ln(2):
```

We plot the corresponding solution curve in the xt-plane with the command

```
> plot(t(x), x = -3..1);
```

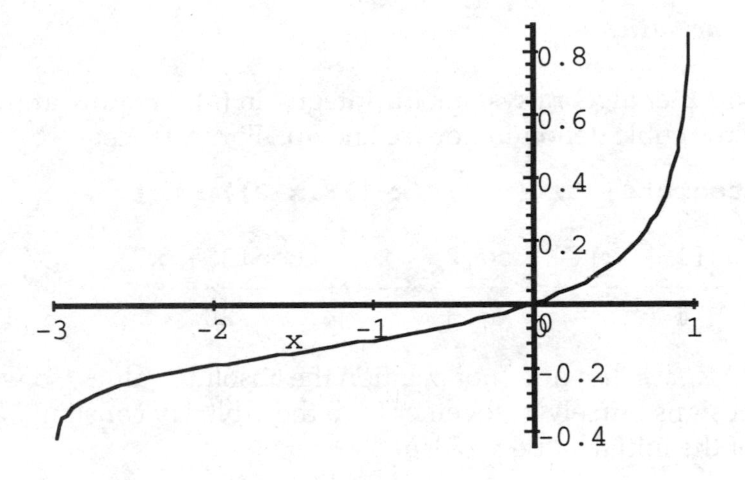

For the differential equation in (10) — where the cubic on the right-hand side does not factor rationally — we get

```
> int(1/(x^3 - 3*x^2 + 1), x);
```

$$\sum_{\_R=\%1} \_R\ln(x+54\_R^2+9\_R-5)$$

$$\%1 := \text{RootOf}(81\_Z^3-9\_Z+1)$$

for the integral in (2). Thus we have a sum of three terms, with the three values of the parameter $\_R$ being the roots of the cubic equation $81z^3 - 9z + 1 = 0$, which are given approximately by

```
> R := fsolve(81*z^3 - 9*z + 1 = 0, z);
```

$$R := -.3790526809,\ .131643615,\ .2474090663$$

Here R is the list R[1], R[2], R[3] of three distinct roots. Substituting these three roots in the sum indicated above, we get the expression

```
> sum(R[i]*ln(x + 54*R[i]^2 + 9*R[i] - 5), i=1..3);
```

$$-.3790526809\ \ln(x-.652703643)+.131643615\ \ln(x-2.879385242)$$
$$+.2474090663\ \ln(x+.532088886)$$

for the value of the integral in (2). Here again, we must supply absolute values and the appropriate value of the arbitrary constant before calculating or plotting.

Using *Mathematica*

With a computer algebra system, the integral in (4) is readily at hand if the zeros a, b, c of the cubic denominator are known. For instance:

```
Integrate[ 1/((x+3)*(x-1)*(x-2)), x ]

-Log[1 - x]     Log[2 - x]     Log[3 + x]
----------- + ---------- + ----------
     4             5             20
```

Note that *Mathematica* does not mention the absolute values, so we must supply the correct signs ourselves, together with the arbitrary constant. For instance, a solution of the initial value problem

$$x' = (x+3)(x-1)(x-2), \qquad x(0) = 0$$

in the open interval (–3, 1) is defined implicitly by

```
t[x_] := (1/20)*Log[x+3]-(1/4)*Log[1-x]+(1/5)*Log[2-x]-
              (1/20)*Log[3]-(1/5)*Log[2]
```

We plot the corresponding solution curve in the xt-plane with the command

```
Plot[t[x], {x,-2.99,0.99}, AxesLabel->{"x","t"} ]
```

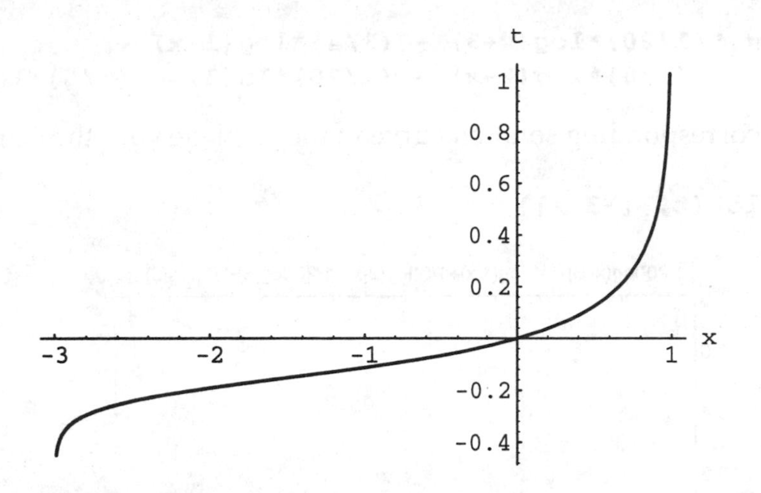

In the case of the differential equation in (10) — where the cubic on the right-hand side does not factor rationally — we first approximate the three roots numerically:

```
R = NSolve[ x^3 - 3*x^2 + 1 == 0, x]
{{x -> -0.532089}, {x -> 0.652704}, {x -> 2.87939}}
a = R[[1,1,2]];   b = R[[2,1,2]];   c = R[[3,1,2]];
```

With these three roots, the integral on the right-hand side in (4) is then given by

```
t = Integrate[ 1/((x-a)(x-b)(x-c)), x]

-0.379053 Log[0.652704-x] + 0.131644 Log[2.87939-x] +
  0.247409 Log[0.532089+x]
```

Using MATLAB

With a computer algebra system, the integral in (4) is readily at hand if the zeros $a,\ b,\ c$ of the cubic denominator are known. For instance:

```
» int('1/((x+3)*(x-1)*(x-2))')
ans =
        1/20*log(x+3)-1/4*log(x-1)+1/5*log(x-2)
```

Note that MATLAB does not mention the absolute values, so we must supply the correct signs ourselves, together with the arbitrary constant. For instance, a solution of the initial value problem

$$x' = (x+3)(x-1)(x-2), \qquad x(0) = 0$$

in the open interval $(-3, 1)$ is defined implicitly by

```
» t = '(1/20)*log(x+3) - (1/4)*log(1-x) +
          (1/5)*log(2-x) - (1/20)*ln(3) - (1/5)*ln(2)'
```

We plot the corresponding solution curve in the *xt*-plane with the command

```
»ezplot(t, [-3 1])
```

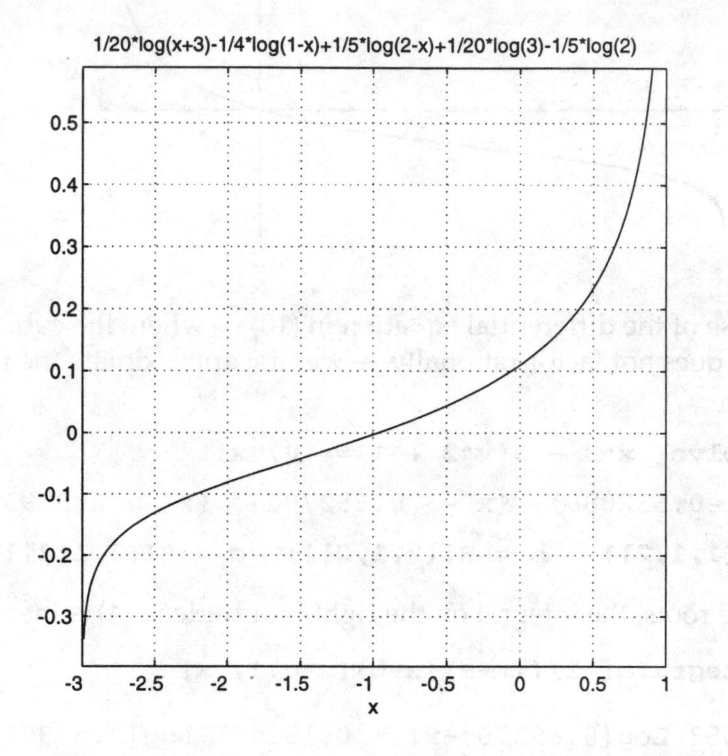

For the differential equation in (10) — where the cubic on the right-hand side does not factor rationally — we get

```
» int('1/(x^3 - 3*x^2 + 1)')
ans =
sum(_R*log(x+54*_R^2+9*_R-5),
            _R = RootOf(81*_Z^3-9*_Z+1))
```

for the integral in (2). Thus we have a sum of three terms, with the three values of the parameter **_R** being the roots of the cubic equation $81z^3 - 9z + 1 = 0$, which are given approximately by

```
» R = numeric(solve('81*z^3 - 9*z + 1 = 0', 'z'))
R =
   0.2474
  -0.3791 + 0.0000i
   0.1316 + 0.0000i
```

Here **R** denotes the is vector **[R(1) R(2) R(3)]** of three distinct roots. With **format long** we see that the imaginary parts of the last two roots appear to represent round-off error, so let's write

```
» R = real(R);
```

Next we substitute 5-digit approximations to these roots in the three symbolic terms of the sum indicated above:

```
» S = 54*R.^2 + 9*R - 5;
» T1 = vpa(symmul(R(1),subs('log(x-s)',S(1),'s')), 5);
» T2 = vpa(symmul(R(2),subs('log(x-s)',S(2),'s')), 5);
» T3 = vpa(symmul(R(3),subs('log(x-s)',S(3),'s')), 5);
```

Finally, the symbolic sum of these three terms,

```
» symop(T1,'+',T2,'+',T3)
ans =
.24741*log(x-.53209)-.37905*log(x+.65270)+
.13164*log(x+2.8794)
```

gives the value of the integral in (2). Here again, we must supply absolute values and the appropriate value of the arbitrary constant before calculating or plotting.

Project 8
Numerical Implementation of Euler's Approximation Method

Reference: Section 2.4 of Edwards & Penney
DIFFERENTIAL EQUATIONS with Computing and Modeling

One's understanding of a numerical algorithm is sharpened by considering its implementation in the form of a calculator or computer program. Figure 2.4.7 in the text lists TI-85 and BASIC programs implementing Euler's method to approximate the solution of the initial value problem

$$\frac{dy}{dx} = x + y, \qquad y(0) = 1 \tag{1}$$

considered in Example 1 of Section 2.5. The comments provided in the final column there should make these programs intelligible even if you have little familiarity with the BASIC and Texas Instruments programming languages. To increase the number of steps (and thereby decrease the step size) you need only change the value of **N** specified in the first line of the program. To apply Euler's method to a different equation $dy/dx = f(x,y)$, you need only change the single line that calculates the function value **F**.

We illustrate below the implementation of Euler's method in systems like *Maple*, *Mathematica*, and MATLAB. For this project, you should implement Euler's method on your calculator or in a programming language of your choice. First test your program by carrying through its application to the initial value problem in (1), and then apply it to solve some of the problems for Section 2.4 in the text. (Euler's method serves mainly to illustrate a fundamental approach to numerical approximate of solutions, and we therefore defer physical applications of approximate solutions to the subsequent projects when more accurate numerical techniques are available.)

Using *Maple*

To apply Euler's method to the initial value problem in (1), we first define the right-hand function $f(x,y) = x + y$ in the differential equation.

```
> f := (x,y) -> x + y;
```

$$f := (x,y) \to x + y$$

To approximate the solution with initial value $y(x_0) = x_0$ on the interval $[x_0, x_f]$, we enter first the initial values

```
> x0 := 0:   y0 := 1:   xf := 1:
```

and then the desired number n of steps and the resulting stepsize h.

```
> n := 10:
> h := evalf((xf - x0)/n);
```

$$h := .10000$$

After we initialize the values of x and y,

```
> x := x0:   y := y0:
```

Euler's method itself is implemented by the following **for** loop, which carries out the iteration

$$y_{n+1} = y_n + h\,f(x_n, y_n), \quad x_{n+1} = x_n + h$$

n times in successtion to take n steps across the interval from $x = x_0$ to $x = x_f$..

```
> for i from 1 to n do
      k := f(x,y):           # the left-hand slope
      y := y + h*k:          # Euler step to update y
      x := x + h:            # update x
      print(x,y);            # display current values
      od:
```

.10000, 1.1000
.20000, 1.2200
.30000, 1.3620
.40000, 1.5282
.50000, 1.7210
.60000, 1.9431
.70000, 2.1974
.80000, 2.4871
.90000, 2.8158
1.0000, 3.1874

Note that x is updated after y in order that the computation $k = f(x, y)$ use the left-hand values (with neither yet updated).

The output consists of x- and y-columns of resulting x_i- and y_i-values. In particular, we see that Euler's method with $n = 10$ steps gives $y(1) \approx 3.1874$ for the initial value problem in (1). The exact solution is $y(x) = 2e^x - x - 1$, so the actual value at $x = 1$ is $y(1) = 2e - 2 \approx 3.4366$. Thus our Euler approximation underestimates the actual value by about 7.25%.

If only the final endpoint result is wanted explicitly, then the print command can be removed from the loop and executed following it:

```
> for i from 1 to n do
      k := f(x,y):           # the left-hand slope
      y := y + h*k:          # Euler step to update y
      x := x + h:            # update x
      od:

> print(x,y);
```

1.0000, 3.1874

For a different initial value problem, we need only enter the appropriate new function $f(x, y)$ and the desired initial and final values in the first two commands above, then re-execute the subsequent ones.

Using *Mathematica*

To apply Euler's method to the initial value problem in (1), we first define the right-hand function $f(x, y) = x + y$ in the differential equation.

```
f[x_,y_] := x + y
```

To approximate the solution with initial value $y(x_0) = x_0$ on the interval $[x_0, x_f]$, we enter first the initial values

```
x0 = 0;       y0 = 1;       xf = 1;
```

and then the desired number n of steps and the resulting stepsize h.

```
n = 10;
h = (xf - x0)/n   //  N
0.1
```

After we initialize the values of x and y,

```
x = x0;     y = y0;
```

Euler's method itself is implemented by the following **Do** loop, which carries out the iteration

$$y_{n+1} = y_n + h\,f(x_n, y_n), \quad x_{n+1} = x_n + h$$

n times in successtion to take n steps across the interval from $x = x_0$ to $x = x_f$..

```
Do[  k = f[x,y];                (* the left-hand slope      *)
     y = y + h*k;               (* Euler step to update y   *)
     x = x + h;                 (* update x                 *)
     Print[x,"        ",y],     (* display current results *)
     {i,1,n} ]

0.1        1.1
0.2        1.22
0.3        1.362
0.4        1.5282
0.5        1.72102
0.6        1.94312
0.7        2.19743
0.8        2.48718
0.9        2.8159
1.         3.18748
```

Note that x is updated after y in order that the computation $k = f(x, y)$ use the left-hand values (with neither yet updated).

The output consists of x- and y-columns of resulting x_i- and y_i-values. In particular, we see that Euler's method with $n = 10$ steps gives $y(1) \approx 3.1875$ for the initial value problem in (1). The exact solution is $y(x) = 2e^x - x - 1$, so the actual value at $x = 1$ is $y(1) = 2e - 2 \approx 3.4366$. Thus our Euler approximation underestimates the actual value by about 7.25%.

If only the final endpoint result is wanted explicitly, then the print command can be removed from the loop and executed following it:

```
Do[  k = f[x,y];               (* the left-hand slope      *)
     y = y + h*k;              (* Euler step to update y   *)
     x = x + h,                (* update x                 *)
     {i,1,n} ]

Print[x,"        ",y]
1.       3.18748
```

For a different initial value problem, we need only enter the appropriate new function $f(x, y)$ and the desired initial and final values in the first two commands above, then re-execute the subsequent ones.

Using MATLAB

To apply Euler's method to the initial value problem in (1), we first define the right-hand function $f(x, y)$ in the differential equation. User-defined functions in MATLAB are defined in (ASCII) text files. To define the function $f(x, y) = x + y$ we save the MATLAB function definition

```
function  yp = f(x,y)
yp = x + y;     % yp = y'
```

in the text file `f.m`.

To approximate the solution with initial value $y(x_0) = x_0$ on the interval $[x_0, x_f]$, we enter first the initial values

```
» x0 = 0;     y0 = 1;     xf = 1;
```

and then the desired number n of steps and the resulting stepsize h.

```
» n = 10;
» h = (xf - x0)/n
h =
    0.1000
```

After we initialize the values of x and y,

```
» x = x0;     y = y0;     X = x;     Y = y;
```

Euler's method itself is implemented by the following **for** loop, which carries out the iteration

$$y_{n+1} = y_n + h f(x_n, y_n), \qquad x_{n+1} = x_n + h$$

n times in successtion to take n steps across the interval from $x = x_0$ to $x = x_f$.

```
» for i = 1 : n                  % for i = 1 to n do
     k = f(x,y);                 % the left-hand slope
     y = y + h*k;                % Euler step to update y
     x = x + h;                  % update x
     X = [X; x];                 % adjoin new x-value
     Y = [Y; y];                 % adjoin new y-value
  end
```

Note that x is updated after y in order that the computation $k = f(x, y)$ use the left-hand values (with neither yet updated).

As output the loop above produces the resulting column vectors **X** and **Y** of x- and y-values that can be displayed simultaneously using the command

```
» [X,Y]
ans =
          0     1.0000
     0.1000     1.1000
     0.2000     1.2200
     0.3000     1.3620
     0.4000     1.5282
     0.5000     1.7210
     0.6000     1.9431
     0.7000     2.1974
     0.8000     2.4872
     0.9000     2.8159
     1.0000     3.1875
```

In particular, we see that $y(1) \approx 3.1875$ for the initial value problem in (1). If only this final endpoint result is wanted explicitly, then we can simply enter

```
» [X(n+1), Y(n+1)]
ans =
     1.0000     3.1875
```

The index **n+1** (instead of **n**) is required because the initial values x_0 and y_0 are the initial vector elements **X(1)** and **Y(1)**, respectively.

The exact solution of the initial value problem in (1) is $y(x) = 2e^x - x - 1$, so the actual value at $x = 1$ is $y(1) = 2e - 2 \approx 3.4366$. Thus our Euler approximation underestimates the actual value by about 7.25%.

For a different initial value problem, we need only define the appropriate function $f(x, y)$ in the file **f.m**, then enter the desired initial and final values in the first command above and re-execute the subsequent ones.

The **for** loop above can be automated by saving the MATLAB function definition

```
function  [X,Y] = euler1(x,xf,y,n)

h = (xf - x)/n;                        % step size
X = x;                                 % initial x
Y = y;                                 % initial y
for i = 1 : n                          % begin loop
    y = y + h*f(x,y);                  % Euler iteration
    x = x + h;                         % new x
    X = [X;x];                         % update x-column
    Y = [Y;y];                         % update y-column
    end                                % end loop
```

in the text file **euler1.m** (we use the name **euler1** to avoid conflict with MATLAB's built-in **euler** function). This function assumes that the function $f(x, y)$ has been defined and saved in the MATLAB file **f.m**.

The function **euler1** applies Euler's method to take n steps from x to x_f starting with the initial value y of the solution. For instance, with **f** as previously defined, the command

```
» [X,Y] = euler1(0,1, 1, 10);
```

is a one-liner that generates the table **[X,Y]** displayed above to approximate the solution of the initial value problem $y' = x + y, \quad y(0) = 1$ on the x-interval $[0, 1]$.

Project 9
The Improved Euler Method

Reference: Section 2.5 of Edwards & Penney
DIFFERENTIAL EQUATIONS with Computing and Modeling

Figure 2.5.11 in the text lists TI-85 and BASIC programs implementing the improved Euler method to approximate the solution of the initial value problem

$$\frac{dy}{dx} = x + y, \qquad y(0) = 1 \tag{1}$$

considered in Example 2 of Section 2.5. The comments provided in the final column should make these programs intelligible even if you have little familiarity with the BASIC and TI programming languages. To apply the improved Euler method to a differential equation $dy/dx = f(x, y)$, one need only change the initial line of the program, in which the function f is defined. To increase the number of steps (and thereby decrease the step size) one need only change the value of N specified in the second line of the program.

We illustrate below the implementation of the improved Euler method in systems like *Maple, Mathematica*, and MATLAB. To begin this project, you should implement the improved Euler method on your calculator or in a programming language of your choice. First test your program by carrying through its application to the initial value problem in (1), and then apply it to solve some of the problems for Section 2.5 in the text. Then carry out the following two investigations.

Investigation A
Here you are to investigate *approximate* solutions of your personal logistic differential equation

$$\frac{dP}{dt} = \frac{1}{n} P(m - P)$$

from Project 1 (that is, m and n are the *largest* two distinct digits in your student ID number). Starting with an initial population $P(0) = 1$, construct a table comparing the Euler and improved Euler approximations (starting with step size $h = 1$, then repeating with smaller step sizes) with the exact population $P(t)$ for the first 10 years thereafter.

Investigation B
This is an investigation of the following two inital value problems:

(a) $y' = y$, $y(0) = 1$, for which $y(1) = e$

(b) $y' = 4/(1 + x^2)$, $y(0) = 0$, for which $y(1) = \pi$

Your objective is to approximate the famous numbers e and π accurate to

- 1 decimal place using Euler's method;
- 3 decimal places using the improved Euler method;

In each case, start with $n = 10$, and then double the number of subintervals successively until the $y(1)$ results with n and with $2n$ agree to the desired number of decimal places.

Using *Maple*

To apply the improved Euler method to the initial value problem in (1), we first define the right-hand function $f(x, y) = x + y$ in the differential equation.

```
> f := (x,y) -> x + y;
```

$$f := (x,y) \to x+y$$

To approximate the solution with initial value $y(x_0) = y_0$ on the interval $[x_0, x_f]$, we enter first the initial values

```
> x0 := 0:   y0 := 1:   xf := 1:
```

and then the desired number n of steps and the resulting stepsize h.

```
> n := 10:
> h := evalf((xf - x0)/n);
```

$$h := .10000$$

After we initialize the values of x and y,

```
> x := x0:   y := y0:
```

the improved Euler method itself is implemented by the following `for` loop, which carries out the iteration

$$\begin{aligned} k_1 &= f(x_n, y_n), & k_2 &= f(x_n + h, y_n + h\,k_1), \\ k &= \tfrac{1}{2}(k_1 + k_2), \\ y_{n+1} &= y_n + h\,k, & x_{n+1} &= x_n + h \end{aligned}$$

n times in successtion to take n steps across the interval from $x = x_0$ to $x = x_f$.

```
> for i from 1 to n do
      k1 := f(x,y):              # the left-hand slope
      k2 := f(x+h,y+h*k1):       # the right-hand slope
      k := (k1 + k2)/2:          # the average slope
      y := y + h*k:              # Euler step to update y
      x := x + h:                # update x
      print(x,y);                # display current values
      od:
```

.100000, 1.11000
.200000, 1.24205
.300000, 1.39847

```
.400000,  1.58181
.500000,  1.79490
.600000,  2.04087
.700000,  2.32316
.800000,  2.64559
.900000,  3.01238
1.00000,  3.42818
```

Note that x is updated after y in order that the computation $k_1 = f(x, y)$ use the left-hand values (with neither yet updated).

The output consists of x- and y-columns of resulting x_i- and y_i-values. In particular, we see that the improved Euler method with $n = 10$ steps gives $y(1) \approx 3.42818$ for the initial value problem in (1). The exact solution is $y(x) = 2e^x - x - 1$, so the actual value at $x = 1$ is $y(1) = 2e - 2 \approx 3.43656$. Thus our improved Euler approximation underestimates the actual value by about 0.24% (as compared with the 7.25% error observed in the Euler approximation of Project 8).

If only the final endpoint result is wanted explicitly, then the print command can be removed from the loop and executed immediately following it (just as we did with the Euler loop in Project 8). For a different initial value problem, we need only enter the appropriate new function $f(x, y)$ and the desired initial and final values in the first two commands above, then re-execute the subsequent ones.

Using *Mathematica*

To apply the improved Euler method to the initial value problem in (1), we first define the right-hand function $f(x, y) = x + y$ in the differential equation.

```
f[x_,y_] := x + y
```

To approximate the solution with initial value $y(x_0) = y_0$ on the interval $[x_0, x_f]$, we enter first the initial values

```
x0 = 0;    y0 = 1;    xf = 1;
```

and then the desired number n of steps and the resulting stepsize h.

```
n = 10;
h = (xf - x0)/n  //  N
0.1
```

after we initialize the values of x and y,

```
x = x0;    y = y0;
```

the improved Euler method itself is implemented by the following **Do** loop, which carries out the iteration

$$k_1 = f(x_n, y_n), \qquad k_2 = f(x_n + h, y_n + h\,k_1),$$
$$k = \tfrac{1}{2}(k_1 + k_2),$$
$$y_{n+1} = y_n + h\,k, \qquad x_{n+1} = x_n + h$$

n times in successtion to take n steps across the interval from $x = x_0$ to $x = x_f$.

```
Do[  k1 = f[x,y];                (* left-hand slope      *)
     k2 = f[x + h, y + h*k1];    (* right-hand slope     *)
     k  = (k1 + k2)/2;           (* average slope        *)
     y = y + h*k;                (* improved Euler step *)
     x = x + h;                  (* update x             *)
     Print[x,"       ",y],       (* display x and y      *)
     {i,1,n} ]

0.1       1.11
0.2       1.24205
0.3       1.39847
0.4       1.5818
0.5       1.79489
0.6       2.04086
0.7       2.32315
0.8       2.64558
0.9       3.01236
1.        3.42816
```

Note that x is updated after y in order that the computation $k_1 = f(x, y)$ use the left-hand values (with neither yet updated).

The output consists of x- and y-columns of resulting x_i- and y_i-values. In particular, we see that the improved Euler method with $n = 10$ steps gives $y(1) \approx 3.42816$ for the initial value problem in (1). The exact solution is $y(x) = 2e^x - x - 1$, so the actual value at $x = 1$ is $y(1) = 2e - 2 \approx 3.43656$. Thus our improved Euler approximation underestimates the actual value by about 0.24% (as compared with the 7.25% error observed in the Euler approximation of Project 8).

If only the final endpoint result is wanted explicitly, then the print command can be removed from the loop and executed immediately following it (just as we did with the Euler loop in Project 8). For a different initial value problem, we need only enter the appropriate new function $f(x, y)$ and the desired initial and final values in the first two commands above, then re-execute the subsequent ones.

Using MATLAB

To apply the improved Euler method to the initial value problem in (1), we first define the right-hand function $f(x, y)$ in the differential equation. User-defined functions in MATLAB are defined in (ASCII) text files. To define the function $f(x, y) = x + y$ we save the MATLAB function definition

```
function  yp = f(x,y)
yp = x + y;      % yp = y'
```

in the text file `f.m`.

To approximate the solution with initial value $y(x_0) = y_0$ on the interval $[x_0, x_f]$, we enter first the initial values

```
» x0 = 0;     y0 = 1;    xf = 1;
```

and then the desired number n of steps and the resulting stepsize h.

```
» n = 10;
» h = (xf - x0)/n
h =
    0.1000
```

After we initialize the values of x and y,

```
» x = x0;     y = y0;     X = x;     Y = y;
```

the improved Euler method itself is implemented by the following `for` loop, which carries out the iteration

$$k_1 = f(x_n, y_n), \qquad k_2 = f(x_n + h, y_n + h\,k_1),$$
$$k = \tfrac{1}{2}(k_1 + k_2),$$
$$y_{n+1} = y_n + h\,k, \qquad x_{n+1} = x_n + h$$

n times in successtion to take n steps across the interval from $x = x_0$ to $x = x_f$.

```
» for i = 1 : n                   % for i = 1 to n do
      k1 = f(x,y);                % left-hand slope
      k2 = f(x+h,y+h*k1);         % right-hand slope
      k = (k1 + k2)/2;            % average slope
      y = y + h*k;                % Euler step to update y
      x = x + h;                  % update x
      X = [X; x];                 % adjoin new x-value
      Y = [Y; y];                 % adjoin new y-value
      end
```

Note that x is updated after y in order that the computation $k = f(x, y)$ use the left-hand values (with neither yet updated).

As output the loop above produces the resulting column vectors **X** and **Y** of x- and y-values that can be displayed simultaneously using the command

```
» [X,Y]
ans =
          0    1.0000
     0.1000    1.1100
     0.2000    1.2421
     0.3000    1.3985
     0.4000    1.5818
     0.5000    1.7949
     0.6000    2.0409
     0.7000    2.3231
     0.8000    2.6456
     0.9000    3.0124
     1.0000    3.4282
```

In particular, we see that $y(1) \approx 3.4282$ for the initial value problem in (1). If only this final endpoint result is wanted explicitly, then we can simply enter

```
» [X(n+1), Y(n+1)]
ans =
     1.0000    3.4282
```

The index **n+1** (instead of **n**) is required because the initial values x_0 and y_0 are the initial vector elements **X(1)** and **Y(1)**, respectively.

The exact solution of the initial value problem in (1) is $y(x) = 2e^x - x - 1$, so the actual value at $x = 1$ is $y(1) = 2e - 2 \approx 3.4366$. Thus our improved Euler approximation underestimates the actual value by about 0.24% (as compared with the 7.25% error observed in the Euler approximation of Project 8).

For a different initial value problem, we need only define the appropriate function $f(x, y)$ in the file **f.m**, then enter the desired initial and final values in the first command above and re-execute the subsequent ones.

The **for** loop above is really a bit long for ready entry in MATLAB's command mode. The following function was defined by simple editing of the function **euler1** of Project 8.

```
function  [X,Y] = impeuler(x,xf,y,n)

h = (xf - x)/n;                    % step size
X = x;                             % initial x
Y = y;                             % initial y
```

```
for i = 1 : n                        % begin loop
    k1 = f(x,y);                     % left-hand slope
    k2 = f(x+h,y+h*k1);              % right-hand slope
    k = (k1 + k2)/2;                 % average slope
    y = y + h*k;                     % improved Euler step
    x = x + h;                       % new x
    X = [X;x];                       % update x-column
    Y = [Y;y];                       % update y-column
    end                              % end loop
```

With this function saved in the text file **impeuler.m**, we need only assume also that the function $f(x, y)$ has been defined and saved in the file **f.m**.

The function **impeuler** applies the improved Euler method to take n steps from x to x_f starting with the initial value y of the solution. For instance, with **f** as previously defined, the command

```
» [X,Y] = impeuler(0,1, 1, 10);
```

is a one-liner that generates the table **[X,Y]** displayed above to approximate the solution of the initial value problem $y' = x + y$, $y(0) = 1$ on the x-interval $[0, 1]$.

Project 10

The Runge-Kutta Method

Reference: Section 2.6 of Edwards & Penney
DIFFERENTIAL EQUATIONS with Computing and Modeling

Figure 2.6.10 in the text lists TI-85 and BASIC programs implementing the Runge-Kutta method to approximate the solution of the initial value problem

$$\frac{dy}{dx} = x + y, \qquad y(0) = 1 \tag{1}$$

considered in Example 1 of Section 2.6. The comments provided in the final column should make these programs intelligible even if you have little familiarity with the BASIC and TI programming languages. To apply the Runge-Kutta method to a differential equation $dy/dx = f(x,y)$, one need only change the initial line of the program, in which the function f is defined. To increase the number of steps (and thereby decrease the step size) one need only change the value of **N** specified in the second line of the program.

We illustrate below the implementation of the Runge-Kutta method in systems like *Maple, Mathematica*, and MATLAB. To begin this project, you should implement the Runge-Kutta method on your calculator or in a programming language of your choice. First test your program by carrying through its application to the initial value problem in (1), and then apply it to solve some of the problems for Section 2.6 in the text. Then carry out the following two investigations.

Investigation A
This is an investigation of the following two inital value problems:

(a) $y' = y, \qquad y(0) = 1, \qquad$ for which $y(1) = e$

(b) $y' = 4/(1 + x^2), \qquad y(0) = 0, \qquad$ for which $y(1) = \pi$

Your objective is to approximate the famous numbers e and π accurate to

- 1 decimal place using Euler's method;
- 3 decimal places using the improved Euler method;
- 5 decimal places using the Runge-Kutta method.

In each case, start with $n = 10$, and then double the number of subintervals successively until the $y(1)$ results with n and with $2n$ agree to the desired number of decimal places. (You may already have done part of this work in Project 9.) When you finish you will *really* know that $e \approx 2.71828$ and $\pi \approx 3.14159$.

Investigation B
Suppose a 128-lb skydiver jumps from an airplane at an initial altitude of 10,000 ft, and that her *downward* velocity v (in ft/sec after t sec) satisfies the initial value problem

$$4\frac{dv}{dt} = 128 - \frac{v^2}{200}, \qquad v(0) = 0.$$

(i) Separate the variables to derive the exact solution

$$v(t) = 160 \tanh\left(\frac{t}{5}\right).$$

(You may need to look up the hyperbolic tangent function in your calculus book.) Find her *terminal velocity* v_T when $v' = 0$ (so her speed is no longer increasing).

(ii) Use Runge-Kutta to solve numerically for $v(t)$ during the first 20 sec of freefall. See whether you get consistent results with step sizes $h = 0.2$ and $h = 0.1$, and construct a brief table comparing these approximate results with the exact solution. How long (to the nearest second) does the skydiver take to reach 99% of her terminal velocity?

Using *Maple*

To apply the Runge-Kutta method to the initial value problem in (1), we first define as usual the right-hand function in the differential equation.

```
> f := (x,y) -> x + y:
```

To approximate the solution with initial value $y(x_0) = y_0$ on the interval $[x_0, x_f]$, we enter first the initial values

```
> x0 := 0:   y0 := 1:   xf := 1:
```

and then the desired number n of steps and the resulting stepsize h.

```
> n := 10:
> h := evalf((xf - x0)/n):
```

After we initialize the values of x and y,

```
> x := x0:   y := y0:
```

the Runge-Kutta method itself is implemented by the following **for** loop, which carries out the iteration

$$\begin{aligned}
k_1 &= f(x_n, y_n),\\
k_2 &= f(x_n + \tfrac{1}{2}h, y_n + \tfrac{1}{2}h\,k_1),\\
k_3 &= f(x_n + \tfrac{1}{2}h, y_n + \tfrac{1}{2}h\,k_2),\\
k_4 &= f(x_n + h, y_n + h\,k_3)\\
k &= \tfrac{1}{6}(k_1 + 2k_2 + 2k_2 + k_4),\\
y_{n+1} &= y_n + h\,k, \qquad\qquad x_{n+1} = x_n + h
\end{aligned}$$

n times in successtion to take n steps across the interval from $x = x_0$ to $x = x_f$.

```
> for i from 1 to n do
      k1 := f(x,y):                # the left-hand slope
      k2 := f(x+h/2,y+h*k1/2):     # 1st midpoint slope
      k3 := f(x+h/2,y+h*k3/2):     # 2nd midpoint slope
      k4 := f(x+h,y+h*k3):         # the right-hand slope
      k := (k1+2*k2+2*k3+k4)/6:    # the average slope
```

```
y := y + h*k:                     # Euler step to update y
x := x + h:                       # update x
print(x,y);                       # display current values
od:
```

```
.1000000, 1.110342
.2000000, 1.242806
.3000000, 1.399718
.4000000, 1.583650
.5000000, 1.797443
.6000000, 2.044238
.7000000, 2.327506
.8000000, 2.651082
.9000000, 3.019206
1.000000, 3.436563
```

Thus the Runge-Kutta method with $n = 10$ steps gives $y(1) \approx 3.436563$ for the initial value problem in (1). The actual value of the exact solution $y(x) = 2e^x - x - 1$, at $x = 1$ is $y(1) = 2e - 2 \approx 3.436564$, so with only 10 steps the Runge-Kutta gives nearly 6-decimal place accuracy!

Using *Mathematica*

To apply the Runge-Kutta method to the initial value problem in (1), we first define as usual the right-hand function in the differential equation.

```
f[x_,y_] := x + y
```

To approximate the solution with initial value $y(x_0) = y_0$ on the interval $[x_0, x_f]$, we enter first the initial values

```
x0 = 0;     y0 = 1;     xf = 1;
```

and then the desired number n of steps and the resulting stepsize h.

```
n = 10;
h = (xf - x0)/n  //  N

0.1
```

After we initialize the values of x and y,

```
x = x0;    y = y0;
```

the Runge-Kutta method itself is implemented by the following `Do` loop, which carries out the iteration

$$
\begin{aligned}
k_1 &= f(x_n, y_n), \\
k_2 &= f(x_n + \tfrac{1}{2}h, y_n + \tfrac{1}{2}h\,k_1), \\
k_3 &= f(x_n + \tfrac{1}{2}h, y_n + \tfrac{1}{2}h\,k_2), \\
k_4 &= f(x_n + h, y_n + h\,k_3) \\
k &= \tfrac{1}{6}(k_1 + 2k_2 + 2k_2 + k_4), \\
y_{n+1} &= y_n + h\,k, \qquad x_{n+1} = x_n + h
\end{aligned}
$$

n times in successtion to take n steps across the interval from $x = x_0$ to $x = x_f$.

```
Do[  k1 = f[x,y];                    (* left-hand slope  *)
     k2 = f[x+h/2, y + h*k1/2];      (* 1st midpt slope  *)
     k3 = f[x+h/2, y + h*k2/2];      (* 2nd midpt slope  *)
     k4 = f[x + h, y + h*k3   ];     (* right-hand slope *)
     k  = (k1 + 2k2 + 2k3 + k4)/6;(* average slope    *)

     y = y + h*k;                    (* Runge-Kutta step *)
     x = x + h;                      (* update x         *)
     Print[x,"          ",y],        (* display x and y  *)
     {i,1,n} ]
```

```
0.1      1.11034
0.2      1.24281
0.3      1.39972
0.4      1.58365
0.5      1.79744
0.6      2.04424
0.7      2.3275
0.8      2.65108
0.9      3.0192
1.       3.43656
```

Thus the Runge-Kutta method with $n = 10$ steps gives $y(1) \approx 3.43656$ for the initial value problem in (1). The actual value of the exact solution $y(x) = 2e^x - x - 1$, at $x = 1$ is $y(1) = 2e - 2 \approx 3.436564$, so with only 10 steps the Runge-Kutta gives 5-decimal place accuracy!

Using MATLAB

To apply the Runge-Kutta method to the initial value problem in (1), suppose as usual that the right-hand side function

```
function  yp = f(x,y)
yp = x + y;      % yp = y'
```

in the differential equation has been defined and saved in the text file **f.m**. Now the iteration

$$
\begin{aligned}
k_1 &= f(x_n, y_n), \\
k_2 &= f(x_n + \tfrac{1}{2}h, y_n + \tfrac{1}{2}h\,k_1), \\
k_3 &= f(x_n + \tfrac{1}{2}h, y_n + \tfrac{1}{2}h\,k_2), \\
k_4 &= f(x_n + h, y_n + h\,k_3) \\
k &= \tfrac{1}{6}(k_1 + 2k_2 + 2k_2 + k_4), \\
y_{n+1} &= y_n + h\,k, \\
x_{n+1} &= x_n + h
\end{aligned}
$$

is a bit too lengthy for comfortable entry in the form of a command-line **for** loop.

The following Runge-Kutta function **rk** was defined by simple editing of the function **impeuler** of Project 9.

```
function   [X,Y] = rk(x,xf,y,n)

h = (xf - x)/n;                   % step size
X = x;                            % initial x
Y = y;                            % initial y
for i = 1 : n                     % begin loop
    k1 = f(x,y);                  % left-hand slope
    k2 = f(x+h/2,y+h*k1/2);       % 1st midpoint slope
    k3 = f(x+h/2,y+h*k2/2);       % 2nd midpoint slope
    k4 = f(x+h,y+h*k3);           % right-hand slope
    k = (k1+2*k2+2*k3+k4)/6;      % average slope
    y = y + h*k;                  % Runge-Kutta step
    x = x + h;                    % new x
    X = [X;x];                    % update x-column
    Y = [Y;y];                    % update y-column
    end                           % end loop
```

With this function saved in the text file **rk.m**, it needs only that the function $f(x, y)$ has been defined and saved in the file **f.m**. The function **rk** takes as input the initial value **x**, the initial value **y**, the final value **xf** of x, and the desired number **n** of subintervals. As output it produces the resulting column vectors **X** and **Y** of x- and y-values. For instance, for $n = 10$ Runge-Kutta steps from $x = 0$ to $x_f = 1$ the MATLAB command

```
»[X,Y] = rk(0, 1,  1,  10);
```

yields the results

```
»[X,Y]
ans =
   0.1    1.11034
   0.2    1.24281
   0.3    1.39972
```

```
0.4    1.58365
0.5    1.79744
0.6    2.04424
0.7    2.32750
0.8    2.65108
0.9    3.01920
1.0    3.43656
```

(where we have rounded off the 14 decimal places produced by **`format long`**).

Thus the Runge-Kutta method with $n = 10$ steps gives $y(1) \approx 3.43656$ for the initial value problem in (1). The actual value of the exact solution $y(x) = 2e^x - x - 1$, at $x = 1$ is $y(1) = 2e - 2 \approx 3.436564$, so with only 10 steps the Runge-Kutta gives 5-decimal place accuracy!

Chapter 3

Linear Equations of Higher Order

Project 11
Solution Curves for Second-Order Linear Equations

Reference: Section 3.1 of Edwards & Penney
DIFFERENTIAL EQUATIONS with Computing and Modeling

With a graphics calculator or a computer graphing program you can construct for yourself a picture like Fig. 3.1.3 in the text, illustrating a family of solution curves of the differential equation

$$y'' + 3y' + 2y = 0 \tag{1}$$

passing through the (same) point $(0, 1)$. Proceed as follows:

- First show that the general solution of Eq. (1) is

$$y(x) = c_1 e^{-x} + c_2 e^{-2x}. \tag{2}$$

- Then show that the general solution of Eq. (1) satisfying the initial conditions $y(0) = a,\ y'(0) = b$ is

$$y(x) = (2a + b)e^{-x} - (a + b)e^{-2x} \tag{3}$$

In the sections following the problems below, we illustrate the use of *Maple*, *Mathematica*, or MATLAB to plot simultaneously a family of solution curves with different slopes, all passing through the same initial point $(0, a)$ in the xy-plane. After using your own computer system to produce your own version of Fig. 3.1.3, construct for each of the following second-order differential equations a variety of typical solution curves passing through a single fixed point such as (0,1).

1. $y'' - y = 0$

2. $y'' - 3y' + 2 = 0$

3. $2y'' + 3y' + y = 0$

4. $y'' + y = 0$ (with general solution $y(x) = c_1 \cos x + c_2 \sin x$)

5. $y'' + 2y' + y = 0$ (with general solution $y(x) = e^{-x}(c_1 \cos x + c_2 \sin x)$)

Using *Maple*

We can derive the general solution in (3) using the command

```
> genSoln :=
  dsolve( {diff(y(x),x,x) + 3*diff(y(x),x) + 2*y(x)=0,
           y(0)=a, D(y)(0)=b}, y(x) ):
```

Then the particular solution with $y(0) = 1, \; y'(0) = b$ is given by

```
> partSoln := subs(a=1,rhs(genSoln));
```

$$partSoln := (-1-b)e^{(-2x)} + (2+b)e^{(-x)}$$

The set of such particular solutions with initial slopes $b = -5, -4, -3, \ldots, 4, 5$ is then defined by

```
> curves := {seq(partSoln, b=-5..5)}:
```

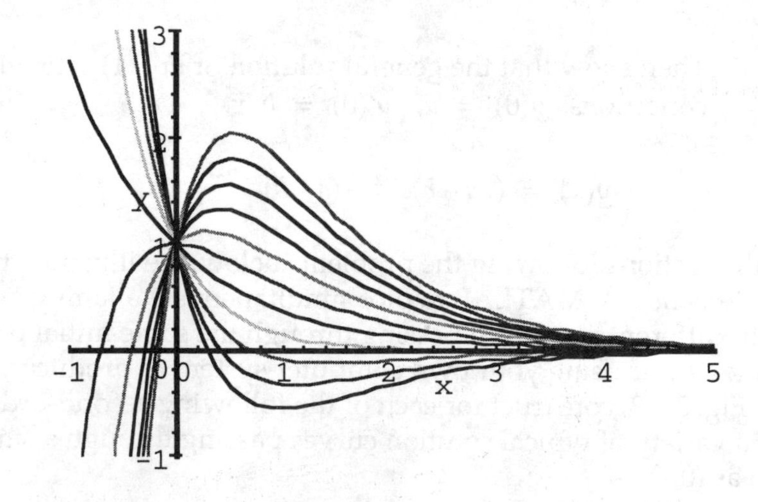

Finally, we plot these 11 curves simultaneously on the x-interval (–1, 5) with the single command

```
> plot( {curves[1], curves[2], curves[3], curves[4],
         curves[5], curves[6], curves[7], curves[8],
         curves[9], curves[10], curves[11]},
         x=-1..5, y=-1..3 );
```

Using *Mathematica*

We can derive the general solution in (3) using the command

```
genSoln = DSolve[ {y''[x] + 3 y'[x] + 2 y[x] == 0,
                   y[0]==a, y'[0]==b}, y[x], x]

             -a - b   2 a + b
{{y[x] -> ------ + -------}}
              2 x       x
             E         E
```

Then the particular solution with $y(0) = 1$, $y'(0) = b$ is given by

```
partSoln = genSoln[[1,1,2]] /. a -> 1

-1 - b   2 + b
------ + -----
  2 x      x
 E        E
```

The set of such particular solutions with initial slopes $b = -5, -4, -3, \ldots, 4, 5$ is then defined by

```
curves = Table[ partSoln, {b,-5,5} ];
```

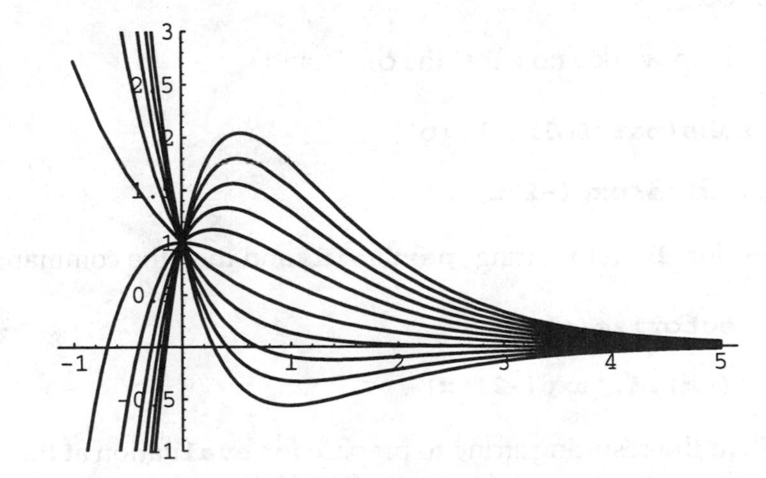

Finally, we plot these 11 curves simultaneously on the x-interval (–1, 5) with the single command

```
Plot[ Evaluate[curves], {x,-1,5}, PlotRange -> {-1,3} ]
```

Using MATLAB

We can derive the general solution in (3) using the command

```
» genSoln = dsolve('D2y + 3*Dy + 2*y,y(0)=a,Dy(0)=b')
genSoln =
(2*a+b)*exp(-x)+(-a-b)*exp(-2*x)
```

Then the particular solution with $y(0) = 1, \; y'(0) = b$ is given by

```
» partSoln = subs(genSoln,1,'a')
partSoln =
(2+b)*exp(-x)+(-1-b)*exp(-2*x)
```

To plot solution curves on the interval $[-1, 5]$ first define the vector

```
» x = -1 : 0.025 : 5;
```

Then the particular solutions with initial slopes $b = -5, -4, -3, \dots, 4, 5$ are plotted by the loop

```
» for b = -5 : 5
     y = subs(partSoln,b,'b');
     y = eval(vectorize(y));
     plot(x,y,'w')
     axis([-1 5 -1 3]), hold on
     end
» grid on
```

To see how this loop works, note that the command

```
» y = subs(partSoln,-5,'b')
y =
-3*exp(-x)+4*exp(-2*x)
```

substitutes `-5` for `b` in the string `partSoln`, and then the command

```
» y = vectorize(y)
y =
-3.*exp(-x)+4.*exp(-2.*x)
```

inserts periods in the resulting string to prepare for **eval**uation at the vector **x**.

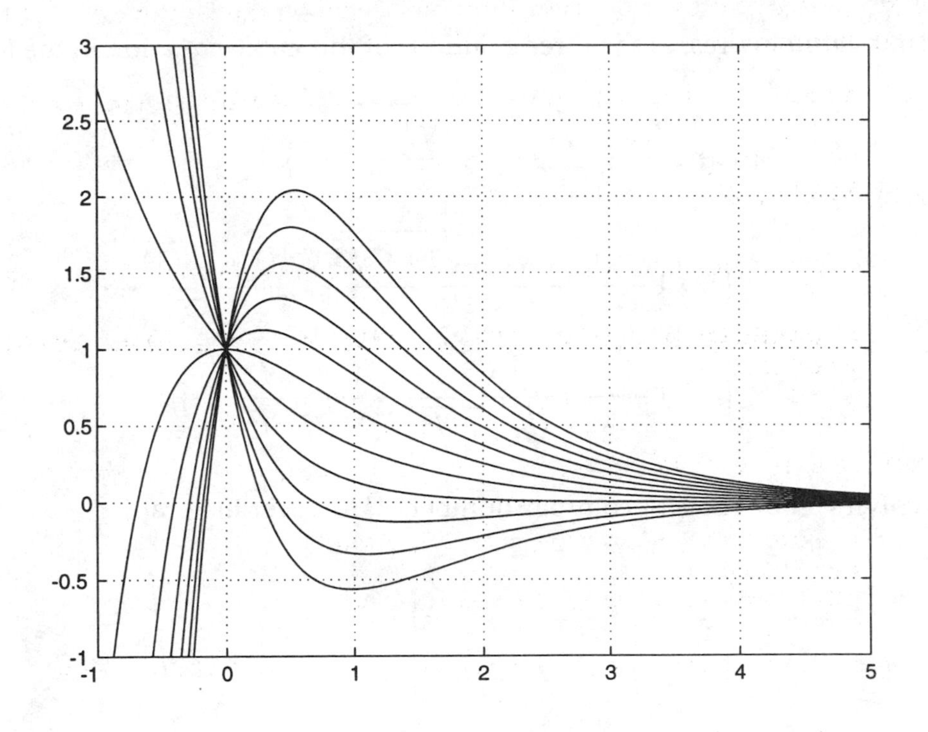

Project 12
Computer Solution of Second-Order Linear Equations

Reference: Section 3.3 of Edwards & Penney
DIFFERENTIAL EQUATIONS with Computing and Modeling

Section 3.3 in the text shows that the solution of a homogeneous linear differential equation reduces to the solution of its characteristic (polynomial) equation. For this and similar purposes, polynomial-solving utilities are now a common feature of calculator and computer systems, and can be used to solve a characteristic equation numerically even when no simple and explicit factorization is evident or even possible. For instance, suppose that we want to solve the homogeneous linear differential equation

$$y''' - 3y'' + y = 0 \tag{1}$$

with characteristic equation

$$r^3 - 3r^2 + 1 = 0. \tag{2}$$

Mathematica expresses the three solutions of this cubic equation in the form

$$\{\{r \to 1 + \frac{\sqrt[3]{2}}{\sqrt[3]{1 + i\sqrt{3}}} + \frac{\sqrt[3]{1 + i\sqrt{3}}}{\sqrt[3]{2}}\},$$

$$\{r \to 1 - \frac{(1 - i\sqrt{3})\sqrt[3]{1 + i\sqrt{3}}}{2\sqrt[3]{2}} - \frac{(1 + i\sqrt{3})^{2/3}}{2^{2/3}}\},$$

$$\{r \to 1 - \frac{1 - i\sqrt{3}}{2^{2/3}\sqrt[3]{1 + i\sqrt{3}}} - \frac{(1 + i\sqrt{3})^{4/3}}{2\sqrt[3]{2}}\}\}$$

involving cube roots of complex numbers. However the graph

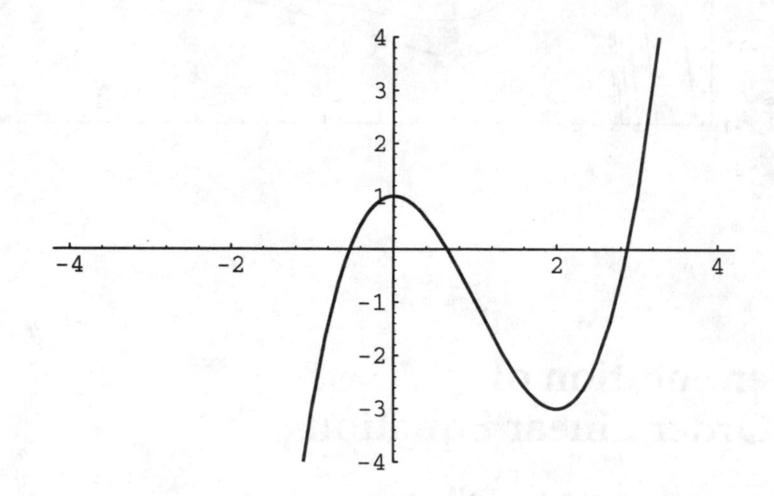

of the function $f(x) = x^3 - 3x^2 + 1$ shows that the three roots of (2) actually are all real numbers. Indeed, with a calculator (like the TI-85) having a built-in polynomial solver, we can simply enter the coefficients $1, -3, 0, 1$ of this cubic polynomial and get the three (approximate) roots $r = -0.5321,\ 0.6527,\ 2.8794$ at the press of a key. Computer systems like *Maple*, *Mathematica*, and MATLAB also have built-in polynomial solvers that can provide such numerical solutions. However we find these roots, it follows that a general solution of the differential equation in (1) is given (approximately) by

$$y(x) = c_1 e^{-0.5321x} + c_2 e^{0.6527x} + c_3 e^{2.8794x}. \tag{3}$$

Computer algebra systems also offer simple DE solver commands for the explicit solution of differential equations. It is interesting to compare the symbolic solutions produced by such DE solvers with explicit numerical solutions of the form in (3).

Use calculator or computer methods to find general solutions (in approximate numerical form) of the following differential equations. Compare the results obtained using a polynomial solver and using a DE solver.

1. $y''' - 3y' + y = 0$

2. $y''' + 3y'' - 3y = 0$

3. $y''' + y' + y = 0$

4. $y''' + 3y' + 5y = 0$

5. $y^{(4)} + 2y''' - 3y = 0$

6. $y^{(4)} + 3y' - 4y = 0$

Using *Maple*

The command

```
> evalf(solve(r^3 - 3*r^2 + 1 = 0, r));
```

— which follows a symbolic solution with a floating-point evaluation — gives the approximate numerical roots $r = -0.5321,\ 0.6527,\ 2.8794$ of the characteristic equation (2) mentioned earlier.

Alternatively, we can first define the differential equation in (1) by entering the command

```
>diffeq := diff(y(x),x$3) - 3*diff(y(x),x$2) + y(x) = 0;
```

$$diffeq := \left(\frac{\partial^3}{\partial^3 x}y(x)\right) - 3\left(\frac{\partial^2}{\partial^2 x}y(x)\right) + y(x) = 0$$

and then ask for its solution by means of the command

```
> dsolve( diffeq, y(x) );
```

$$y(x) = \_C1\ e^{(2\cos(1/9\,\pi)+1)x} + \_C2\ e^{(-\cos(1/9\,\pi) - 1 + \sqrt{3}\ \sin(1/9\,\pi))x} + \_C3\ e^{(-\cos(1/9\,\pi) - 1 - \sqrt{3}\ \sin(1/9\,\pi))x}$$

Evidently *Maple V* (Release 2) has solved the cubic characteristic equation exactly in terms of trigonometric functions! To compare this result with the solution given in (3), we can apply the function **evalf** to convert exact expressions to floating-point approximations:

```
> evalf(");
```

$$y(x) = \_C1\, e^{-0.5321x} + \_C2\, e^{0.6527x} + \_C3\, e^{2.8794x}$$

By comparison, Release 3 of *Maple V* gives the solution in the explicit form

```
> dsolve( diffeq, y(x) );
```

$$y(x) = \sum_{\_R=\%1} \_C1_{\_R}\, e^{(\_R\, x)}$$

$$\%1 := \text{RootOf}(1 - 3\,\_Z^2 + \_Z^3)$$

of a linear combination in terms of the roots of the characteristic equation in (2). An application of **evalf** to this symbolic result returns us again to the numerical approximation in (3).

Using *Mathematica*

The polynomial solver command

```
Solve[ r^3 - 3r^2 + 1 == 0, r]
```

yields the complex "Cardan formula expressions" for the three roots exhibited previously. But the numerical command

```
NSolve[ r^3 - 3r^2 + 1 == 0, r]

{{r -> -0.532089}, {r -> 0.652704}, {r -> 2.87939}}
```

gives the approximate numerical roots $r = -0.5321,\ 0.6527,\ 2.8794$ of the characteristic equation (2) mentioned earlier.

Alternatively, we can first define the differential equation in (1) by entering the command

```
diffeq = y'''[x] - 3 y''[x] + y[x] == 0

                        (3)
y[x] - 3 y''[x] + y        [x] == 0
```

and then ask for its solution by means of the command

```
DSolve[ diffeq, y[x], x ]
```

This gives a complex (and complicated) form of the solution similar to the original symbolic solution of the characteristic equation, but the numerical evaluation

```
DSolve[ diffeq, y[x], x ] // N
                        (2.87939 + 0. I) x
{{y[x] -> 2.71828                             C[1] +

                                         -20
                   (-0.532089 + 5.42101 10     I) x
      2.71828                                          C[2] +

                                        -19
                   (0.652704 - 1.6263 10     I) x
      2.71828                                        C[3]}}
```

of the result gives the approximate numeric form (3) of the solution (with some complex roundoff error) — provided that we recognize the three appearances of the exponential base $e \approx 2.71828$.

Using MATLAB

We can work either in a purely numeric mode or in a symbolic mode. For a numerical approach, the characteristic polynomial in (2) is defined by the vector **[1 -3 0 1]** listing its coefficients in order of descending powers. Then (after **format long**) the command

```
» roots([1 -3 0 1])
ans =
   2.87938524157182
   0.65270364466614
  -0.53208888623796
```

yields the three approximate characteristic roots that appear in the general solution (3). The symbolic command

```
» roots = solve('x^3 - 3*x^2 + 1 = 0')
```

yields "Cardan formula expressions" for these three roots similar to the *Mathematica* expressions exhibited originally. But then the command

```
» numeric(roots)
ans =
   2.87938524157182
  -0.53208888623796
   0.65270364466614
```

reproduces the numeric roots obtained previously.

Alternatively, we can first define the differential equation in (1) by entering the symbolic command

```
» diffeq = 'D3y - 3*D2y + y = 0'
diffeq =
D3y - 3*D2y + y = 0
```

and then ask for its solution by means of the command

```
» dsolve(diffeq)
ans =
C1*exp((2*cos(1/9*pi)+1)*x)+
C2*exp(-(cos(1/9*pi)-1+3^(1/2)*sin(1/9*pi))*x)+
C3*exp(-(cos(1/9*pi)-1-3^(1/2)*sin(1/9*pi))*x)

»pretty

y(x) = _C1 exp((2 cos(1/9 pi) + 1) x)

                                          1/2
        + _C2 exp(- (cos(1/9 pi) - 1 + 3     sin(1/9 pi)) x)

                                          1/2
        + _C3 exp(- (cos(1/9 pi) - 1 - 3     sin(1/9 pi)) x)
```

Here the MATLAB Symbolic Math Toolbox evidently has called on its *Maple* kernel to solve the characteristic equation exactly in terms of trigonometric functions! But the numeric conversions

```
» numeric(2*cos(1/9*pi)+1)
ans =
   2.87938524157182

» numeric(-(cos(1/9*pi)-1+3^(1/2)*sin(1/9*pi)))
ans =
  -0.53208888623796
» numeric(-(cos(1/9*pi)-1-3^(1/2)*sin(1/9*pi)))
ans =
   0.65270364466614
```

finally verify that the symbolic solution agrees with the approximate solution (3) corresponding to our numerical solution of the characteristic equation.

Project 13

Computer Algebra Implementation of the Method of Undetermined Coefficients

Reference: Section 3.5 of Edwards & Penney
DIFFERENTIAL EQUATIONS with Computing and Modeling

The use of a computer algebra system can lighten the burden of algebraic computation associated with the method of undetermined coefficients. The paragraphs below illustrate the use of *Maple*, *Mathematica*, and MATLAB to find particular solutions of non-homogeneous differential equations by this method of "educated guessing". Whichever system you are using, it might be instructive to peruse the similar implementations of undetermined coefficients in different systems. Then you will be ready to apply the method to differential equations — such as those of Problems 21 through 30 in Section 3.5 of the text — where a manual approach might be impractical. We will defer physical applications of undetermined coefficients to Project 15.

Using *Maple*

We want to apply the method of undetermined coefficients to find a particular solution of the nonhomogeneous differential equation

```
> deq := 7*diff(y(x),x$3) + 4*diff(y(x),x$2) -
              5*diff(y(x),x) + 12*y(x)   =   3516*sin(3*x);
```

$$7\left(\frac{\partial^3}{\partial x^3}y(x)\right)+4\left(\frac{\partial^2}{\partial x^2}y(x)\right)-5\left(\frac{\partial}{\partial x}y(x)\right)+12y(x) = 3156\sin(3\,x).$$

It is not necessary to solve the associated homogeneous equation to find the complementary function. We need only substitute sin(3 x) in the right-hand side of **deq**,

```
> eval(subs(y(x)=sin(3*x),lhs(deq)));
```

$$-204\cos(3\,x)-24\sin(3\,x)$$

The nonzero result shows that there is no duplication between the complementary function and the trial solution

```
> yp := A*cos(3*x) + B*sin(3*x);
```

Substitute of this trial solution into our differential equation gives

```
> eq:= eval(subs(y(x)=yp, deq));
```

$$eq1 := 204\,A\sin(3\,x) - 204\,B\cos(3\,x) - 24\,A\cos(3\,x) - 24\,B\sin(3\,x) = 3516\sin(3\,x)$$

We can now equate coefficients of sines and cosines and solve for A and B.

```
> solve({-24*A-204*B=0, 204*A-24*B=3516}, {A,B});
```

$$\{B = -2, A = 17\}$$

Thus our particular solution is defined by

```
> partSoln := subs(A=17,B=-2,yp);
```

$$partSoln := 17\cos(3\,x) - 2\sin(3\,x)$$

You might well ask we don't simple apply *Maple*'s **dsolve** function to **deq**. Try it and see! Perhaps you then will agree that it's simpler to carry out the method of undetermined coefficients as illustrated above.

Using *Mathematica*

We want to apply the method of undetermined coefficients to find a particular solution of the non-homogeneous differential equation

$$L[y] = y^{(5)} + 2y^{(4)} + 2y^{(3)} - 300x^2 - 30e^x = 0 \qquad (1)$$

with complementary function

$$y_c = A + Bx + Cx^2 + e^{-x}(C\cos x + E\sin x).$$

Noting the duplication between the complementary function and the non-homogeneous term $300x^2$, we first specify the form

```
y = a x^3 + b x^4 + c x^5 + d E^x;
```

of the desired trial solution. We then calculate the result of substituting **y** in the left-hand side of (1) with the command

```
Ly = D[y,{x,5}] + 2 D[y,{x,4}]+
                  2 D[y,{x,3}]- 300 x^2 - 30 E^x
```

```
              x         x            2                     x
120 c - 30 E   + d E   - 300 x   + 2 (24 b + d E   + 120 c x) +

             x                  2
   2 (6 a + d E  + 24 b x + 60 c x )
```

where `D[y,{x,n}]` yields the *n*th derivative of the expression `y` with respect to the variable `x`. Upon collecting coefficients of the the types of terms that appear,

```
Collect[ Ly, {1,x,x^2,E^x} ]
```

```
                                             x
12 a + 48 b + 120 c + (-30 + 5 d) E   + (48 b + 240 c) x +

                      2
   (-300 + 120 c) x
```

we can readily read off the linear equations to be solved for the coefficients *a, b, c, d.*

```
coeffs = Solve[ { 12 a + 48 b + 120 c == 0,
                   -30 + 5 d == 0,
                  (48 b + 240 c) == 0,
                  -300 + 120c == 0},   {a,b,c,d} ]
```

```
                                    25         5
{{a -> 25, d -> 6, b -> -(--), c -> -}}
                                    2          2
```

Thus the desired particular solution of Equation (1) is given by

```
partSoln = yp == y /. coeffs[[1]]
```

```
                                4       5
          x           3   25 x      5 x
yp == 6 E   + 25 x   - ----- + ----
                            2       2
```

You might well ask we don't simple apply *Mathematica*'s **DSolve** function to the differential equation in (1). Try it and see! Perhaps you then will agree that it's simpler to carry out the method of undetermined coefficients as illustrated above.

Using MATLAB

We want to apply the method of undetermined coefficients to find a particular solution of the non-homogeneous differential equation

$$y''' - 3y'' - 4y' + 12y \ = \ 375\, x^2 e^{3x} \qquad (2)$$

with complementary function

$$y_c = c_1 e^{-2x} + c_2 e^{2x} + c_3 e^{3x}.$$

Noting the duplication between the complementary function and the nonhomogeneous term $375\,x^2 e^{3x}$, we first specify the appropriate form

```
» yp = '(A*x + B*x^2 + C*x^3)*exp(3*x)'
yp =
(A*x + B*x^2 + C*x^3)*exp(3*x)
```

of the desired trial solution. We then calculate the result of substituting **yp** into Equation (2) — rewritten with the nonhomogeneous term on the left — using the symbolic command

```
» symop(diff(yp,3),'-','3*',diff(yp,2),'-
','4*',diff(yp),
          '+','12*',yp,'-','375*x^2*exp(3*x)')
ans =
6*C*exp(3*x)+6*(2*B+6*C*x)*exp(3*x)+
5*(A+2*B*x+3*C*x^2)*exp(3*x)-375*x^2*exp(3*x)
```

After collecting coefficients of like powers of x,

```
» collect
ans =
(15*C*exp(3*x)375*exp(3*x))*x^2+
(36*C*exp(3*x)+10*B*exp(3*x))*x+
6*C*exp(3*x)+12*B*exp(3*x)+5*A*exp(3*x)
```

we can easily read off the equations that the coefficients A, B, C must satisfy in order that the result reduce to 0.

```
» [A,B,C] = solve('15*C-375=0',
                  '36*C+10*B=0',
                  '6*C+12*B+5*A=0', 'A,B,C')
A = 186
B = -90
C = 25
```

We finally obtain our particular solution of Equation (2) by substituting these values of the coefficients in **yp**.

```
» yp = subs(yp,A,'A'); yp = subs(yp,B,'B');
  yp = subs(yp,C,'C')
yp =
(186*x-90*x^2+25*x^3)*exp(3*x)
```

Can you see that this result is consistent with the general solution of (2) given by

```
» dsolve 'D3y - 3*D2y - 4*Dy + 12*y = 375*x^2*exp(3*x)'
ans =
-90*x^2*exp(3*x)+186*x*exp(3*x)-936/5*exp(3*x)
+25*x^3*exp(3*x)+
C1*exp(2*x)+C2*exp(3*x)+C3*exp(-2*x)
```

Or is there an "extra term" here?

Project 14
Computer Algebra Implementation of the Method of Variation of Parameters

Reference: Section 3.5 of Edwards & Penney
DIFFERENTIAL EQUATIONS with Computing and Modeling

The method of variation of parameters — as described in the final part of Section 3.5 in the text — is readily implemented using a computer algebra system. In the paragraphs below we illustrate the use of *Maple*, *Mathematica*, and MATLAB in finding a particular solution of the differential equation

$$y'' + y = \tan x \tag{1}$$

of Example 11, to which the method of undetermined coefficients does not apply (why?). In each case the first three lines of code enter the two independent homogeneous solutions $y_1 = \cos x$ and $y_2 = \sin x$ and the nonhomogeneous term $f(x) = \tan x$ in Equation (1). In order to apply the method of variation of parameters to a different nonhomogeneous differential equation, we would need only to alter these three initial lines appropriately to fit the new equation, and then re-execute the remaining lines of code. Try this with Problems 47-62 in Section 3.5.

Using *Maple*

As indicated above, we first enter

```
> y1 := cos(x):               # independent
  y2 := sin(x):               # complementary solutions
  f  := tan(x):               # nonhomogeneous function
```

We next set up and solve the linear pair of equations in (31) in the text.

```
> y1p := diff(y1, x):
  y2p := diff(y2, x):        # Derivatives of y1 and y2

  eqs := { u1p*y1  +  u2p*y2  = 0,
           u1p*y1p +  u2p*y2p = f}:

  soln := solve( eqs, {u1p, u2p} );
```

$$soln := \left\{u2p = \frac{\cos(x)\tan(x)}{\sin(x)^2 + \cos(x)^2}, u1p = -\frac{\tan(x)\sin(x)}{\sin(x)^2 + \cos(x)^2}\right\}$$

```
> u1p := rhs(soln[2]):
  u2p := rhs(soln[1]):
```

Finally we integrate **u1p** = u_1' and **u2p** = u_2' and assemble the desired particular solution $y_p = u_1 y_1 + u_2 y_2$.

```
> u1 := int(u1p, x):
  u2 := int(u2p, x):
   y := simplify( u1*y1 + u2*y2);
```

$$y := \ln(1-\cos(x)-\sin(x))\cos(x) - \ln(1-\cos(x)+\sin(x))\cos(x) - \sin(x)$$

By combining the two logarithms on the right and then converting sines and cosines to secants and tangents, you should be able to verify that this solution is equivalent to the one found manually in the text.

Using *Mathematica*

As indicated in the introduction to this project, we first enter

```
y1  = Cos[x];          (* independent              *)
y2  = Sin[x];          (* complementary solutions  *)
f   = Tan[x];          (* nonhomogeneous function  *)
```

We next set up and solve the linear pair of equations in (31) in the text.

```
y1p = D[y1,x];    y2p = D[y2,x];   (* the derivatives *)

eqs = {u1p y1  + u2p y2  == 0,
       u1p y1p + u2p y2p == f};

soln = Solve[ eqs, {u1p,u2p} ]

                  Sin[x] Tan[x]                        Sin[x]
{{u1p ->-(------------------), u2p -> -----------------}}
                    2         2                   2         2
              Cos[x]  + Sin[x]              Cos[x]  + Sin[x]
```

```
u1p = soln[[1,1,2]];
u2p = soln[[1,2,2]];     (* pick out the derivatives *)
```

Finally we integrate **u1p** = u_1' and **u2p** = u_2' and assemble the desired particular solution $y_p = u_1 y_1 + u_2 y_2$.

```
u1 = Integrate[u1p,x];
u2 = Integrate[u2p,x];

y = Simplify[ u1 y1 + u2 y2 ]

                    x           x                 x          x
Cos[x] (Log[Cos[-] - Sin[-]] - Log[Cos[-] + Sin[-]])
                    2           2                 2          2
```

The familiar trigonometric identities

$$\sin x = 2\sin x/2\,\cos x/2 \quad \text{and} \quad \cos x = \cos^2 x/2 - \sin^2 x/2$$

can be used to verify that this solution is equivalent to the one found manually in the text.

Using MATLAB

As indicated in the introduction to this project, we first enter

```
» y1 = 'cos(x)';        % independent
» y2 = 'sin(x)';        % complementary solutions
» f  = 'tan(x)';        % nonhomogeneous function
```

We next set up and solve the linear pair of equations in (31) in the text.

```
» y1p = diff(y1);
» y2p = diff(y2);     % Derivatives of y1 and y2

» eq1 = symop('u1p','*',y1,'+','u2p','*',y2,'=0')
eq1 =
u1p*cos(x)+u2p*sin(x) = 0

» eq2 = symop('u1p','*',y1p,'+','u2p','*',y2p,'=',f)
eq2 =
-u1p*sin(x)+u2p*cos(x) = tan(x)

» [u1p,u2p] = solve(eq1,eq2,'u1p,u2p')
u1p =
-1/(sin(x)^2+cos(x)^2)*tan(x)*sin(x)
u2p =
1/(sin(x)^2+cos(x)^2)*cos(x)*tan(x)

» pretty
```

```
                tan(x) sin(x)                 cos(x) tan(x)
   {u1p = - -----------------, u2p = -----------------}
                     2         2              2         2
               sin(x)   + cos(x)       sin(x)   + cos(x)
```

Finally we integrate **u1p** = u_1' and **u2p** = u_2' and assemble the desired particular solution $y_p = u_1 y_1 + u_2 y_2$.

```
» u1 = int(u1p);
» u2 = int(u2p);
» y = simplify(symop(u1,'*',y1,'+',u2,'*',y2))
y =
log(1-cos(x)-sin(x))*cos(x)-
     log(1-cos(x)+sin(x))*cos(x)-sin(x)
```

By combining the two logarithms on the right and then converting sines and cosines to secants and tangents, you should be able to verify that this solution is equivalent to the one found manually in the text.

Project 15

Steady Periodic and Transient Solutions

Reference: Section 3.6 of Edwards & Penney *DIFFERENTIAL EQUATIONS with Computing and Modeling*

For your personal project let $p = 2a$ and $q = a^2 + b^2$, where a and b are the two smallest nonzero digits of your student I.D. number (with $a < b$), and let r be the largest such digit. Then solve the initial value problem

$$\begin{aligned} x'' + p\,x' + q\,x &= r\cos t, \\ x(0) = x_0, \quad x'(0) &= v_0 \end{aligned} \tag{1}$$

in terms of x_0 and v_0. With a computer algebra system you can simply apply the "dsolve" command to solve this initial value problem. However, we recommend the hybrid approach of first noting that the complementary function of the differential equation in (1) takes the form of the transient solution

$$x_{tr}(t) = e^{-at}(c_1 \cos bt + c_2 \sin bt) \tag{2}$$

(why?), and then applying the method of undetermined coefficients as in Project 13 to find the values of the coefficients A and B in the steady periodic (particular) solution

$$x_{sp}(t) = A\cos t + B\sin t. \tag{3}$$

Finally, determine the values of the coefficients c_1 and c_2 in (2) in terms of the initial position and velocity x_0 and v_0. Then generate a figure like Fig. 3.8.9 in the text, except showing solution curves with $x_0 = 0$ and various different values of v_0.

Note that the external "forcing frequency" in (1) is $\omega = 1$. For a more general investigation, consider the differential equation

$$x'' + p\,x' + q\,x \;=\; r \cos \omega t, \tag{4}$$

and express the coefficients A and B in the steady periodic (particular) solution

$$x_{sp}(t) \;=\; A \cos \omega t + B \sin \omega t \tag{5}$$

in terms of the external frequency ω. It should be interesting to graph the amplitude $C = \sqrt{A^2 + B^2}$ of $x_{sp}(t)$. This graph should exhibit a "peak" that corresponds to the resonance frequency of the system.

Automobile Vibrations

The upward displacement function $x(t)$ of the car of Example 5 in the text satisfies the differential equation

$$m\,x'' + c\,x' + k\,x \;=\; c\,y' + k\,y \tag{6}$$

when the car's shock absorber is connected (so $c > 0$). With $y = a \sin \omega t$ for the road surface, this equation becomes

$$m\,x'' + c\,x' + k\,x \;=\; E_0 \cos \omega t + F_0 \sin \omega t \tag{7}$$

where $E_0 = c\omega a$ and $F_0 = ka$. Apply the result of Problem 22 in this section to show that the amplitude C of the resulting steady periodic oscillations of the car is given by

$$C \;=\; \frac{a\sqrt{k^2 + (c\omega)^2}}{\sqrt{(k - m\omega^2)^2 + (c\omega)^2}} \tag{8}$$

Because $\omega = 2\pi v / L$ when the car is moving with velocity v, this gives C as a function of v.

Using the numerical data given in Example 5 in the text (including c = 3000 N-s/m), see whether you can use Eq. (8) to reproduce the plot in Fig. 3.8.12 showing the amplitude of the car's vibrations as a function of its velocity over the washboard surface. The graph indicates that, as the car accelerates gradually from rest, it initially oscillates with amplitude slightly over 5 cm. Maximum resonance oscillations with amplitude about 14 cm occur around 32 mi/h, but then subside to more tolerable levels at high speeds. Can you verify the maximum point on the graph by maximizing C as a function of ω in Eq. (8)?

The paragraphs below illustrate *Maple*, *Mathematica*, and MATLAB techniques that will be useful in carrying out the investigations of this project. For this purpose we will use the forced mass-spring-dashpot system

$$m\,x'' + c\,x' + k\,x = F_0 \cos \omega t \tag{9}$$

with the numerical parameter values $m = 25$, $c = 10$, $k = 226$, and $F_0 = 100$.

Using *Maple*

We start with the mass-spring-dashpot parameters

```
> m := 25:    c := 10:    k := 226:
```

and the associated homogeneous differential equation

```
> de1 := m*diff(x(t),t,t)+c*diff(x(t),t)+k*x(t) = 0:
```

whose (transient) solution is given by

```
> dsolve(de1, x(t));
```

$$x(t) = \_C1\,e^{-t/5}\cos(3t) + \_C2\,e^{-t/5}\sin(3t)$$

We see that the natural frequency is $\omega_0 = 3$. We now ask what forcing frequency ω produces the steady periodic response of maximal amplitude. The (nonhomogeneous) differential equation in (9) is

```
> de2 := m*diff(x(t),t,t)+c*diff(x(t),t)+k*x(t)
         = F0*cos(w*t):
```

To carry out the method of undetermined coefficients, we set up the trial function

```
> trial := A*cos(w*t) + B*sin(w*t):
```

and substitute it into **de2**:

```
> eval(subs(x(t)=trial, de2)):
> collect(", [cos(w*t),sin(w*t)]);
```

$$(-25\,A\,w^2 + 10\,B\,w + 226\,A)\cos(wt)$$
$$+(-25\,B\,w^2 - 10\,A\,w + 226\,B)\sin(wt) \;=\; F0\cos(wt)$$

Equating coefficients of **cos(wt)** and of **sin(wt)**, we get two equations to be solved for **A** and **B**.

```
> soln := solve({-25*A*w^2 + 10*B*w + 226*A = F0,
                 -25*B*w^2 -10*A*w + 226*B = 0},
                 {A,B} ):
> A := rhs(soln[2]);   B := rhs(soln[1]);
```

$$A := -\frac{(25w^2 - 226)\,F0}{-11200\;w^2 + 625\,w^4 + 51076}$$

$$B := 10\frac{w\,F0}{-11200\;w^2 + 625\,w^4 + 51076}$$

Substituting $F_0 = 100$, the square of the amplitude of the response function is then given by

```
> F0 := 100:       Csq := simplify(A^2 + B^2):
```

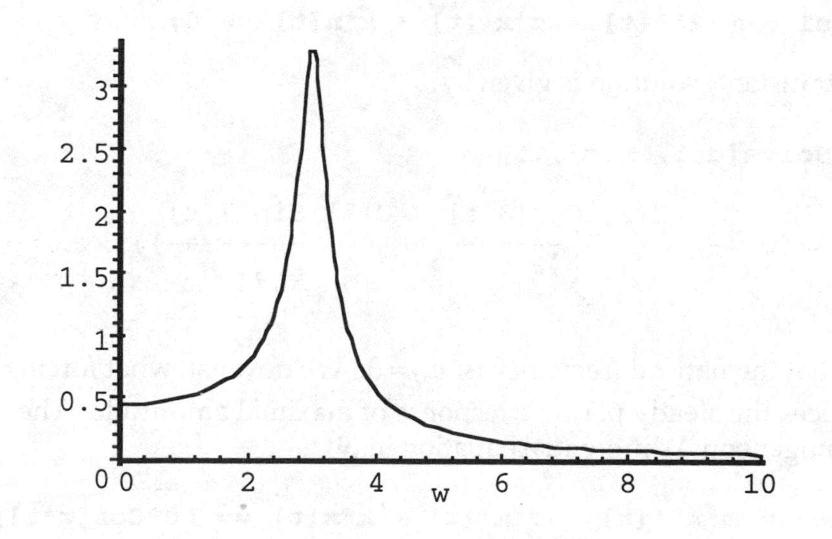

When we plot C as a function of ω, we see a sharp peak indicating "practical resonance" near the natural frequency of $\omega_0 = 3$.

```
> plot(sqrt(Csq), w = 0..10);
```

The actual maximum value of the response amplitude will occur where the derivative is zero.

```
> diff(Csq, w);
```

$$-10000 \frac{-22400\, w + 2500\, w^3}{(-11200\, w^2 + 625\, w^4 + 51076)^2}$$

```
> solve(-22400 + 2500*w^2, w):
> evalf(");
```

$$2.993326, \; -2.993326$$

Thus the resonance frequency of the system is given by $\omega \approx 2.993326$ (just a bit less than $\omega_0 = 3$).

Using *Mathematica*

We start with the mass-spring-dashpot parameters

```
m = 25;    c = 10;    k = 226;
```

and the associated homogeneous differential equation

```
de1 = m*x''[t] + c*x'[t] + k*x[t] == 0;
```

whose (transient) solution is given by

```
DSolve[de1, x[t], t]

                C[2] Cos[3 t]    C[1] Sin[3 t]
{{x[t] -> ------------- - -------------}}
                    t/5              t/5
                   E                E
```

We see that the natural frequency is $\omega_0 = 3$. We now ask what forcing frequency ω produces the steady periodic response of maximal amplitude. The (nonhomogeneous) differential equation in (9) is

```
de2 = m*x''[t] + c*x'[t] + k*x[t] == F0*Cos[w*t];
```

To carry out the method of undetermined coefficients, we set up the trial function

```
trial[t_] := A Cos[w t] + B Sin[w t]
```

and substitute it into **de2**:

```
de2 /.   x  ->  trial
rhs = Collect[First[%], {Cos[w t],Sin[w t]} ]
```

```
                             2
(226 A + 10 B w - 25 A w ) Cos[t w] +

                               2
  (226 B - 10 A w - 25 B w ) Sin[t w]
```

Equating coefficients of **Cos[w*t]** and of **Sin[w*t]**, we get two equations to be solved for **A** and **B**.

```
soln = Solve[ {Coefficient[rhs, Cos[w t]] == F0,
                Coefficient[rhs, Sin[w t]] == 0 },
              {A,B} ]
```

```
                                     2
                     F0 (226 - 25 w )
{{A -> -(---------------------------),
                               2        4
            -51076 + 11200 w   - 625 w

                    -10 F0 w
    B -> ---------------------------}}
                               2        4
            -51076 + 11200 w   - 625 w
```

```
A = First[A /. soln];
B = First[B /. soln];
```

Substituting $F_0 = 100$, the square of the amplitude of the response function is then given by

```
F0 = 10;
Csq = Simplify[A^2 + B^2];
```

When we plot C as a function of ω, we see a sharp peak indicating "practical resonance" near the natural frequency of $\omega_0 = 3$.

```
Plot[ Sqrt[Csq], {w,0,10} ]
```

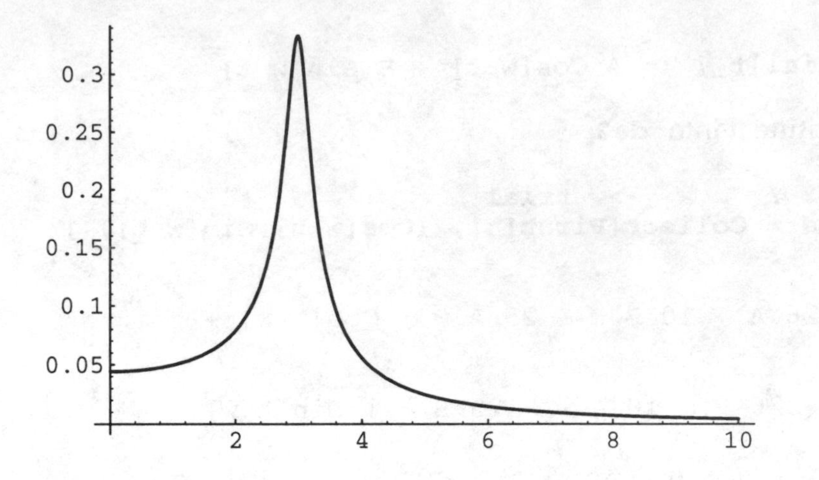

The actual maximum value of the response amplitude will occur where the derivative is zero.

```
D[Csq, w]
```

```
                              3
 -100 (-22400 w + 2500 w )
-----------------------------
                  2          4 2
(51076 - 11200 w  + 625 w )
```

```
NSolve[ -22400 + 2500 w^2 == 0, w ]
```

```
{{w -> -2.99333}, {w -> 2.99333}}
```

Thus the resonance frequency of the system is given by $\omega \approx 2.99333$ (just a bit less than $\omega_0 = 3$).

Using MATLAB

We start with the homogeneous differential equation

```
» de1 = '25*D2x + 10*Dx + 226*x = 0';
```

associated with Eq. (9) (substituting the given parameter values $m = 25$, $c = 10$, and $k = 226$). Its (transient) solution is given by

```
» dsolve(de1)
ans =
C1*exp(-1/5*t)*sin(3*t)+C2*exp(-1/5*t)*cos(3*t)
```

We see that the natural frequency is $\omega_0 = 3$. We now ask what forcing frequency ω produces the steady periodic response of maximal amplitude. The (nonhomogeneous) differential equation in (9) is

```
» de2 = '25*D2x + 10*Dx + 226*x = 100*cos(w*t)';
```

(with $F_0 = 100$ substituted). To carry out the method of undetermined coefficients, we set up the trial function

```
»  xp = 'A*cos(w*t) + B*sin(w*t)';
```

and substitute it into **de2**:

```
» sub =
symop('25*',diff(xp,'t',2),'+','10*',diff(xp,'t'),
                    '+','226*',xp,'-','100*cos(w*t)');
```

The commands

```
» collect(sub, 'cos(w*t)')
ans =
(-25*A*w^2+10*B*w+226*A-100)*cos(w*t)-
25*B*sin(w*t)*w^2-10*A*sin(w*t)*w+226*B*sin(w*t)

» collect(sub, 'sin(w*t)')
ans =
(-25*B*w^2+226*B-10*A*w)*sin(w*t)-
25*A*cos(w*t)*w^2-
100*cos(w*t)+10*B*cos(w*t)*w+226*A*cos(w*t)
```

then collect the coefficients of **cos(w*t)** and of **sin(w*t)** for us. Thus we get two equations to be solved for **A** and **B**.

```
» [A,B] = solve('-25*A*w^2+10*B*w+226*A = 100',
                    '-25*B*w^2+226*B-10*A*w = 0', 'A,B')
A =
-100*(25*w^2-226)/(-11200*w^2+625*w^4+51076)
B =
1000*w/(-11200*w^2+625*w^4+51076)
```

The square of the amplitude of the response function is then given by

```
» Csq = symop(A,'*',A,'+',B,'*',B);
```

When we plot C as a function of ω, we see a sharp peak indicating "practical resonance" near the natural frequency of $\omega_0 = 3$.

```
» w = 0 : .05 : 10;
» C = sqrt(eval(vectorize(Csq)));
» plot(w,C)
```

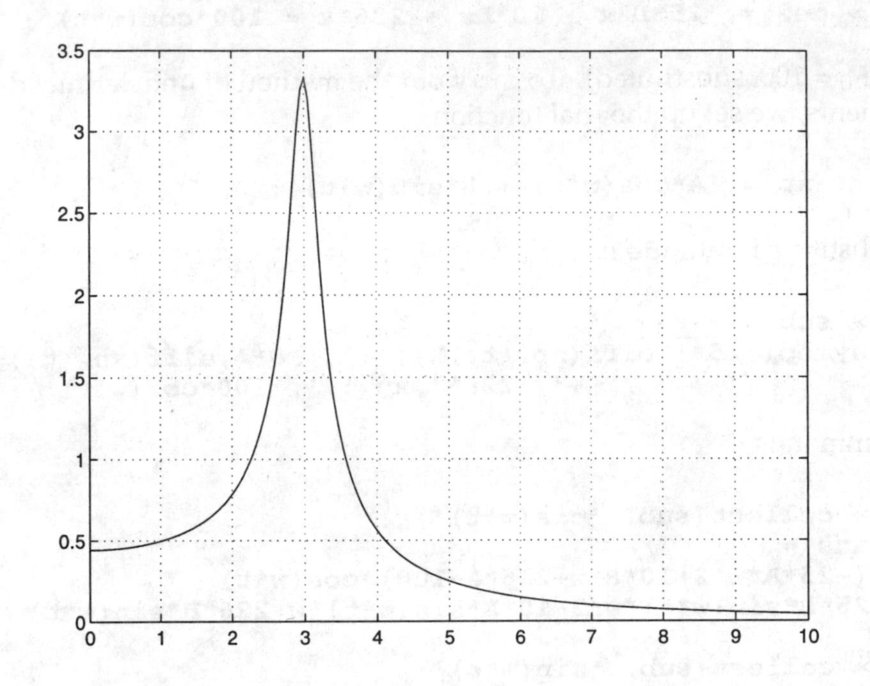

The actual maximum value of the response amplitude will occur where the derivative is zero.

```
» diff(Csq,'w');
» simplify
ans =
-1000000*w*(-224+25*w^2)/(-11200*w^2+625*w^4+51076)^2

» numeric(solve('-224 + 25*w^2'))
ans =
    2.9933
   -2.9933
```

Thus the resonance frequency of the system is given by $\omega \approx 2.9933$ (just a bit less than $\omega_0 = 3$).

Chapter 4

Introduction to Systems of Differential Equations

Project 16
Computer Algebra Solutions of Systems

Reference: Section 4.2 of Edwards & Penney
DIFFERENTIAL EQUATIONS with Computing and Modeling

Computer algebra systems can be used to solve systems as well as single differential equations. The paragraphs below illustrate the use of *Maple*, *Mathematica*, and MATLAB to solve symbolically the systems

$$x' = 4x - 3y, \qquad y' = 6x - 7y \tag{1}$$

of Example 1 and

$$x'' + 3x - y = 0, \qquad y'' - 2x + 2y = 0 \tag{2}$$

of Example 3 in Section 4.2 of the text. Use these examples as models for the computer algebra solution of several of the systems in Problems 1 through 20 in Section 4.2. In each case, you should verify that the general solution obtained by computer is equivalent to the one obtained manually. Frequently the two presumably equivalent solutions will look different at first glance, and you will then need to explore the relation between the arbitrary constants in the computer algebra solution and the arbitrary constants in the manual solution.

Using *Maple*

To solve the system in (1) we need only define the differential equations

```
> deq1 := diff(x(t),t) = 4*x(t) - 3*y(t);
  deq2 := diff(y(t),t) = 6*x(t) - 7*y(t);
```

$$deq1 := \frac{\partial}{\partial x} x(t) = 4x(t) - 3y(t)$$

$$deq2 := \frac{\partial}{\partial x} y(t) = 6x(t) - 7y(t)$$

Then the command

```
> dsolve( {deq1,deq2}, {x(t),y(t)} );
```

$$\left\{x(t) = \frac{3}{2}\_C1\, e^{2t} + \_C2\, e^{-5t},\ y(t) = \_C1\, e^{2t} + 3\_C2\, e^{-5t}\right\}$$

yields the same general solution found in the text using the method of elimination (though with the second arbitrary constant differing by a factor of 3), and the command

```
> dsolve( {deq1,deq2, x(0)=2,y(0)=-1}, {x(t),y(t)} );
```

$$\left\{y(t) = 2\, e^{2t} - 3\, e^{-5t},\ x(t) = 3\, e^{2t} - e^{-5t}\right\}$$

solves the initial value problem of Example 1 in the text. For the second-order system in (2) we use the commands

```
>  deq3 := diff(x(t),t,t) + 3*x(t) -   y(t) = 0:
   deq4 := diff(y(t),t,t) - 2*x(t) + 2*y(t) = 0:

> dsolve( {deq3,deq4}, {x(t),y(t)} );
```

$$\left\{\begin{array}{l} y(t) = 2\_C1\sin(t) + 2\_C2\cos(t) - \_C3\sin(2t) - \_C4\cos(2t), \\ x(t) = \_C1\sin(t) + \_C2\cos(t) + \_C3\sin(2t) + \_C4\cos(2t) \end{array}\right\}$$

that yield the same general solution found in Example 3 in the text.

Using *Mathematica*

To solve the system in (1) we need only define the differential equations

```
deq1 =       x'[t] == 4 x[t] - 3 y[t]
deq2 =       y'[t] == 6 x[t] - 7 y[t]

x'[t] == 4 x[t] - 3 y[t]
y'[t] == 6 x[t] - 7 y[t]
```

Then the command

```
DSolve[{deq1,deq2}, {x[t],y[t]}, t ]

                              2 t                              2 t
                -2         9 E                  3           3 E
{{x[t] -> (------ + ------) C[1] + (------ - ------) C[2],
                5 t        7                  5 t          7
             7 E                            7 E
```

```
                 2 t                               2 t
       -6     6 E                     9          2 E
y[t] -> (------ + ------) C[1] + (------ - ------) C[2]}}
        5 t      7                   5 t        7
      7 E                          7 E
```

yields a solution that appears to differ from the general solution

$$x(t) = \frac{3}{2} a_1 e^{2t} + a_2 e^{-5t}, \qquad y(t) = a_1 e^{2t} + 3 a_2 e^{-5t}$$

found in the text using the method of elimination. But upon equating coefficients of e^{2t} and e^{-5t} in $x(t)$ and $y(t)$, we find that the two computations

```
Solve[{(9/7)c1 - (3/7)c2 == (3/2)a1,
       (-2/7)c1 + (3/7)c2 == a2}, {c1,c2} ]
```

and

```
Solve[{(6/7)c1 - (2/7)c2 == a1,
       (-6/7)c1 + (9/7)c2 == 3 a2}, {c1,c2} ]
```

both yield (upon simplification) the same result

```
          3 a1
{{c1 -> ---- + a2, c2 -> a1 + 3 a2}}
           2
```

This verifies the equivalence of the two solutions. Finally, the commands

```
DSolve[{deq1,deq2, x[0]==2,y[0]==-1},
                    {x[t],y[t]}, t ];
Simplify[%]
```

```
              -5 t       2 t               -3         2 t
{{x[t] -> -E       + 3 E    , y[t] -> ---- + 2 E    }}
                                       5 t
                                      E
```

solve the initial value problem of Example 1 in the text. For the second-order system in (2) we define

```
deq3 = D[x[t],t,t] + 3*x[t] -   y[t] == 0;
deq4 = D[y[t],t,t] - 2*x[t] + 2*y[t] == 0;
```

The command

```
genSoln = DSolve[ {deq3,deq4}, {x[t],y[t]}, t];
```

then yields $x(t)$ and $y(t)$ as rather miserable-looking linear combinations of the complex exponentials e^{ix}, e^{-ix}, e^{2ix}, and e^{-2ix} (try it for yourself and see). But we can convert these complex exponentials to trigonometric expressions by means of the commands

```
x = genSoln[[1,1,2]] // ComplexExpand;
Collect[x, {Cos[t],Sin[t],Cos[2t],Sin[2t]} ]]
```

```
           2 I                 I              2 Re[C[1]]     Re[C[2]]
Cos[2 t](--- Im[C[1]] - - Im[C[2]] + ---------- - --------) +
           3                   3                  3              3

          I                 I               Re[C[1]]     Re[C[2]]
  Cos[t]  (- Im[C[1]] + - Im[C[2]] + -------- + --------) +
          3                 3                 3              3

   I               I               Re[C[3]]     Re[C[4]]
  (- Im[C[3]] + - Im[C[4]] + -------- + --------) Sin[t] +
   3               3                 3              3

   I               I               Re[C[3]]     Re[C[4]]
  (- Im[C[3]] - - Im[C[4]] + -------- - --------) Sin[2 t]
   3               6                 3              6
```

```
y = genSoln[[1,2,2]] // ComplexExpand;
Collect[y, {Cos[t],Sin[t],Cos[2t],Sin[2t]} ]
```

```
          -2 I                 I              2 Re[C[1]]     Re[C[2]]
Cos[2t](---- Im[C[1]] + - Im[C[2]] - ---------- + --------) +
           3                   3                  3              3

            2 I                 2 I              2 Re[C[1]]
  Cos[t]  (--- Im[C[1]] + --- Im[C[2]] + ---------- +
            3                   3                  3

    2 Re[C[2]]        2 I                 2 I               2 Re[C[3]]
    ----------) + (--- Im[C[3]] + --- Im[C[4]] + ---------- +
        3               3                   3                   3

   2 Re[C[4]]              -I               I               Re[C[3]]
   ----------) Sin[t] + (-- Im[C[3]] + - Im[C[4]] - -------- +
       3                       3               6                 3

      Re[C[4]]
      --------) Sin[2 t]
         6
```

Upon comparing the coefficients of $\cos(x)$, $\sin(x)$, $\cos(2x)$, and $\sin(2x)$ in these expressions for x and y, we see finally that — with

a_1 = `(Re[C[3]] + I Im[C[3]] + Re[C[4]] + I Im[C[4]])/3`

and similar expressions for a_2, a_3, and a_4 — our solution can be written in the form

$$x(t) = a_1 \sin(t) + a_2 \cos(t) + a_3 \sin(2t) + a_4 \cos(2t),$$
$$y(t) = 2\, a_1 \sin(t) + 2\, a_2 \cos(t) - a_3 \sin(2t) - a_4 \cos(2t)$$

that agrees with the Example 3 solution found in the text.

Using MATLAB

To solve the system in (1) we need only define the differential equations

```
» deq1 =   'Dx = 4*x - 3*y',   deq2 =   'Dy = 6*x - 7*y'
deq1 =
Dx = 4*x - 3*y
deq2 =
Dy = 6*x - 7*y
```

Then the command

```
» [x,y] = dsolve(deq1,deq2,'t')
x =
3/2*exp(2*t)*C2+exp(-5*t)*C1
y =
exp(2*t)*C2+3*exp(-5*t)*C1
```

yields a solution that agrees with the general solution

$$x(t) = \frac{3}{2}\, a_1\, e^{2t} + a_2\, e^{-5t}, \qquad y(t) = a_1\, e^{2t} + 3\, a_2\, e^{-5t}$$

found in the text using the method of elimination (though with the two arbitrary constants interchanged and one of them differing by a factor of 3). The command

```
» [x,y] = dsolve(deq1,deq2,'x(0)=2','y(0)=-1','t')
x =
3*exp(2*t)-exp(-5*t)
y =
2*exp(2*t)-3*exp(-5*t)
```

solves the initial value problem of Example 1 in the text. For the second-order system in (2) we use the commands

```
» deq3 = 'D2x + 3*x - y = 0',
  deq4 = 'D2y - 2*x + 2*y = 0'
```

```
» [x,y] = dsolve(deq3,deq4,'t')
x =
C1*sin(t)+C2*cos(t)-C4*sin(2*t)+C3*cos(2*t)
y =
2*C1*sin(t)+2*C2*cos(t)+C4*sin(2*t)-C3*cos(2*t)
```

that yield the same general solution (except for the order and signs of the four arbitrary constants) as that found in Example 3 in the text.

Project 17

The Runge-Kutta Method for 2-Dimensional Systems

Reference: Section 4.3 of Edwards & Penney *DIFFERENTIAL EQUATIONS with Computing and Modeling*

Figure 4.3.11 in the text lists TI-85 and BASIC versions of the two-dimensional Runge-Kutta program RK2DIM. You should note that it closely parallels the one-dimensional Runge-Kutta program listed in Fig. 2.6.11, with a single line there replaced (where appropriate) with two lines here to calculate a *pair* of x- and y-values or slopes. Note also that the notation used is essentially that of Eqs. (13) and (14) in Section 4.3. The first several lines define the functions and initial data needed for Example 1.

The paragraphs below illustrate the use of *Maple*, *Mathematica*, and MATLAB to apply the Runge-Kutta method to the initial value problem

$$\begin{aligned} x' &= -\pi y, & x(0) &= 1 \\ y' &= \pi x, & y(0) &= 0 \end{aligned} \qquad (1)$$

whose exact solution is given by

$$x(t) = \cos(\pi t), \qquad y(t) = \sin(\pi t).$$

This exercise should prepare you for the following investigations.

Investigation A

Let a be the largest and b the smallest nonzero digit of your student ID number. Then use the Runge-Kutta method to approximate the solution of the initial value problem defined by

$$x' = -a y, \qquad y' = b x$$

with $x(0)$ being the next smallest digit and $y(0)$ the next largest digit of your ID. Also use your system's "dsolve" function to find the exact symbolic solution. Finally, compare your approximate solution points with the corresponding exact solution points.

Investigation B

Suppose that *you* jump from an airplane at an initial altitude of 10,000 ft, and that your *downward* position $x(t)$ and velocity $y(t)$ (in ft/sec after t seconds) satisfy the initial value problem

$$x' = y, \qquad x(0) = 0$$
$$(W/g)y' = W - (0.0015)(100y + y^2), \qquad y(0) = 0$$

where W denotes *your* weight in pounds and $g = 32$ ft/sec^2. Use the Runge-Kutta method to solve numerically for $x(t)$ and $y(t)$ during your first 15 to 20 seconds of freefall, with successive step sizes h and $h/2$ small enough to get results consistent to 2 decimal places. How far have you free-fallen after 10 seconds? After 20 seconds? Use your results (for the first 20 seconds) to determine the total time required for your descent to the ground.

Investigation C

Suppose that a lunar lander is free-falling toward the moon's surface at a speed of 1000 miles per hour. Its retrorockets, when fired in free space, provide a deceleration of 33,000 mi/hr^2. Assume that the radius of the moon is exactly 1080 miles, and that the retrorockets are fired at an altitude of 24.64 miles above the lunar surface. If we measure distance in *kilomiles* and time in *hours*, then the rocket's distance $x(t)$ from the moon's *center* and its velocity $y(t)$ satisfy the initial value problem

$$x' = y, \qquad x(0) = 1.10464$$

$$y' = 33 - \frac{15.1632}{x^2}, \qquad y(0) = -1.$$

Use the Runge-Kutta method to track the lander's descent to the lunar surface where $x = 1.080$ (getting x and y accurate to several decimal places by successively doubling the step size h appropriately). How long (in minutes and seconds) does it take? With what velocity does the lander hit the surface of the moon? Did the pilot choose the right instant to hit the retrorockets? In how deep a crater must it land to achieve a soft touchdown? (For this last question you presumably want to focus on the instant at which $y(0) = 0$.)

Using *Maple*

To apply the Runge-Kutta method to the initial value problem in (1), we let

```
> Digits := 6:  pi := evalf(Pi)  # numerical value of Pi
```

and define the right-hand side functions in our two differential equations:

```
> f := (t,x,y) -> -pi*y:
  g := (t,x,y) ->  pi*x:
```

Suppose that we want to approximate the solution functions $x = \cos(\pi t)$ and $y = \sin(\pi t)$ on the interval $[0, 1/2]$. To approximate the solution with initial values $x(t_0) = x_0$, $y(t_0) = y_0$ on the interval $[t_0, t_f]$, we enter first the initial values

```
> t0 := 0:  x0 := 1:  y0 := 0:  tf := 0.5:
```

and then the desired number n of steps, the interval at which we want to print results, and the resulting stepsize h.

```
> n := 12:                  # number of subintervals
  m := 2:                   # to print every mth step
  h := evalf((tf - t0)/n):  # step size
```

After we initialize the values of t, x, and y,

```
> t := t0:    x := x0:    y := y0:
```

the Runge-Kutta method itself is implemented by the `for` loop below, which carries out the iteration

$$
\begin{aligned}
k_1 &= f(t_n, x_n, y_n), \\
l_1 &= g(t_n, x_n, y_n) \\
k_2 &= f(t_n + \tfrac{1}{2}h, x_n + \tfrac{1}{2}h\,k_1, y_n + \tfrac{1}{2}h\,l_1), \\
l_2 &= g(t_n + \tfrac{1}{2}h, x_n + \tfrac{1}{2}h\,k_1, y_n + \tfrac{1}{2}h\,l_1) \\
k_3 &= f(t_n + \tfrac{1}{2}h, x_n + \tfrac{1}{2}h\,k_2, y_n + \tfrac{1}{2}h\,l_2), \\
l_3 &= g(t_n + \tfrac{1}{2}h, x_n + \tfrac{1}{2}h\,k_2, y_n + \tfrac{1}{2}h\,l_2) \\
k_4 &= f(t_n + h, x_n + h\,k_3, y_n + h\,l_3), \\
l_4 &= f(t_n + h, x_n + h\,k_3, y_n + h\,l_3) \\
k &= \tfrac{1}{6}(k_1 + 2k_2 + 2k_3 + k_4), \\
l &= \tfrac{1}{6}(l_1 + 2l_2 + 2l_3 + l_4) \\
x_{n+1} &= x_n + h\,k, \quad y_{n+1} = y_n + h\,l, \quad t_{n+1} = t_n + h
\end{aligned}
$$

n times in successtion to take n steps across the interval from $t = t_0$ to $t = t_f$. It simplies the printing to define in advance the "formatting string"

```
> fmt := `%3.2f %5.4f %5.4f \n`:
```

where the notation **%w.df** specifies printing the corresponding value a total of w digits and d decimal places. Then Runge-Kutta loop is then

```
> for i from 1 to n do
     k1 := f(t,x,y):
     l1 := g(t,x,y):                          # left-hand slopes
     k2 := f(t+h/2,x+h*k1/2,y+h*l1/2):
     l2 := g(t+h/2,x+h*k1/2,y+h*l1/2):   # 1st midpt slopes
     k3 := f(t+h/2,x+h*k2/2,y+h*l2/2):
     l3 := g(t+h/2,x+h*k2/2,y+h*l2/2):   # 2nd midpt slopes
     k4 := f(t+h,x+h*k3,y+h*l3):
     l4 := g(t+h,x+h*k3,y+h*l3):          # right-hand slopes
     k := (k1+2*k2+2*k3+k4)/6:            # average x-slope
     l := (l1+2*l2+2*l3+l4)/6:            # average y-slope
     x := x + h*k:                   # Euler step to update x
     y := y + h*l:                   # Euler step to update y
     t := t + h:                     # update t
     if trunc(i/m) =i/m then                    # display every
             printf(fmt,180*t,x,y) fi;          #     mth value
     od:
```

```
                    15  0.9659  0.2588
                    30  0.8660  0.5000
                    45  0.7071  0.7071
                    60  0.5000  0.8660
                    75  0.2588  0.9659
                    90  0.0000  1.0000
```

Note that we have arranged to print angles in *degrees* in the first column of output. The second and third columns give the corresponding approximate values of the cosine and sine functions. Noting the familiar values $\cos 60° = \sin 30° = \frac{1}{2}$, $\cos 30° = \sin 60° = \frac{1}{2}\sqrt{3} \approx 0.8660$, and $\cos 45° = \sin 45° = \frac{1}{2}\sqrt{2} \approx 0.7071$, we see that the Runge-Kutta method with only $n = 12$ subintervals has provided 4 decimal places of accuracy on the whole range from 0° to 90°.

If only the final endpoint result is wanted explicitly, then the print command can be removed from the loop and executed immediately following it (just as we did with the Euler loop in Project 8). For a different initial value problem, we need only enter the appropriate functions $f(x,y)$ and $g(x,y)$ for our new differential equations and the desired initial and final values in the initial commands above, then re-execute the subsequent ones.

Using *Mathematica*

To apply the Runge-Kutta method to the initial value problem in (1), we let

```
pi = N[Pi];                        (* numerical value of Pi *)
```

and define the right-hand side functions in our two differential equations:

```
f[t_,x_,y_] := -pi*y
g[t_,x_,y_] :=  pi*x
```

Suppose that we want to approximate the solution functions $x = \cos(\pi t)$ and $y = \sin(\pi t)$ on the interval $[0, 1/2]$. To approximate the solution with initial values $x(t_0) = x_0$, $y(t_0) = y_0$ on the interval $[t_0, t_f]$, we enter first the initial values

```
t0 = 0;    x0 = 1;    y0 = 0;    tf = 0.5;
```

and then the desired number n of steps, the interval at which we want to print results, and the resulting stepsize h.

```
n = 12;               (* number of subintervals  *)
m = 2;                (* to print every mth step *)
h = (tf - t0)/n;      (* step size               *)
```

After we initialize the values of t, x, and y,

```
t = t0;     x = x0;     y = y0;
```

the Runge-Kutta method itself is implemented by the `Do` loop below, which carries out the iteration

$$
\begin{aligned}
k_1 &= f(t_n, x_n, y_n), \\
l_1 &= g(t_n, x_n, y_n) \\
k_2 &= f(t_n + \tfrac{1}{2}h, x_n + \tfrac{1}{2}h\,k_1, y_n + \tfrac{1}{2}h\,l_1), \\
l_2 &= g(t_n + \tfrac{1}{2}h, x_n + \tfrac{1}{2}h\,k_1, y_n + \tfrac{1}{2}h\,l_1) \\
k_3 &= f(t_n + \tfrac{1}{2}h, x_n + \tfrac{1}{2}h\,k_2, y_n + \tfrac{1}{2}h\,l_2), \\
l_3 &= g(t_n + \tfrac{1}{2}h, x_n + \tfrac{1}{2}h\,k_2, y_n + \tfrac{1}{2}h\,l_2) \\
k_4 &= f(t_n + h, x_n + h\,k_3, y_n + h\,l_3), \\
l_4 &= f(t_n + h, x_n + h\,k_3, y_n + h\,l_3) \\
k &= \tfrac{1}{6}(k_1 + 2k_2 + 2k_3 + k_4), \\
l &= \tfrac{1}{6}(l_1 + 2l_2 + 2l_3 + l_4) \\
x_{n+1} &= x_n + h\,k, \qquad y_{n+1} = y_n + h\,l, \qquad t_{n+1} = t_n + h
\end{aligned}
$$

n times in successtion to take n steps across the interval from $t = t_0$ to $t = t_f$.

```
Do[
   k1 = f[t,x,y];
   l1 = g[t,x,y];                          (* left-hand slopes *)
   k2 = f[t+h/2,x+h*k1/2,y+h*l1/2];
   l2 = g[t+h/2,x+h*k1/2,y+h*l1/2]; (* 1st midpt slopes *)
   k3 = f[t+h/2,x+h*k2/2,y+h*l2/2];
   l3 = g[t+h/2,x+h*k2/2,y+h*l2/2]; (* 2nd midpt slopes *)
   k4 = f[t+h,x+h*k3,y+h*l3];
   l4 = g[t+h,x+h*k3,y+h*l3];          (* right-hand slopes *)
   k  = (k1+2*k2+2*k3+k4)/6;            (* average x-slope  *)
   l  = (l1+2*l2+2*l3+l4)/6;            (* average y-slope  *)
   x  = x + h*k;                     (* Euler step to update x  *)
   y  = y + h*l;                     (* Euler step to update y  *)
   t  = t + h;                       (* update t                *)
   If[ Floor[i/m] == i/m,
       Print[180*t,PaddedForm[x,{7,4}],
       PaddedForm[y,{7,4}]]], (* display every mth value *)
   {i,1,n} ]
```

```
15.    0.9659    0.2588
30.    0.8660    0.5000
45.    0.7071    0.7071
60.    0.5000    0.8660
75.    0.2588    0.9659
90.    0.0000    1.0000
```

Note that we have arranged to print angles in *degrees* in the first column of output. The second and third columns give the corresponding approximate values of the cosine and sine functions. The **PaddedForm[x,{7,4}]** specification formats a result by printing it in a "field" 7 spaces width with 4 decimal places. Noting the familiar values $\cos 60° = \sin 30° = \frac{1}{2}$, $\cos 30° = \sin 60° = \frac{1}{2}\sqrt{3} \approx 0.8660$, and $\cos 45° = \sin 45° = \frac{1}{2}\sqrt{2} \approx 0.7071$, we see that the Runge-Kutta method with only $n = 12$ subintervals has provided 4 decimal places of accuracy on the whole range from $0°$ to $90°$.

If only the final endpoint result is wanted explicitly, then the print command can be removed from the loop and executed immediately following it (just as we did with the Euler loop in Project 8). For a different initial value problem, we need only enter the appropriate functions $f(x,y)$ and $g(x,y)$ for our new differential equations and the desired initial and final values in the initial commands above, then re-execute the subsequent ones.

Using MATLAB

To apply the Runge-Kutta method to the initial value problem in (1), we begin by defining the right-hand side functions in our two differential equations:

```
function  z = f(t,x,y)
z = -pi*y;

function  z = g(t,x,y)
z = pi*x;
```

Suppose we want to approximate the solution functions $x = \cos(\pi t)$ and $y = \sin(\pi t)$ on the interval $[0, 1/2]$. To approximate the solution with initial values $x(t_0) = x_0$, $y(t_0) = y_0$ on the interval $[t_0, t_f]$, we enter first the initial values

```
» t0 = 0;     x0 = 1;     y0 = 0;     tf = 0.5;
```

and then the desired number n of steps, the interval at which we want to print results, and the resulting stepsize h.

```
» n = 12;                       % number of subintervals
» m =  2;                       % to print every mth step
» h = (tf - t0)/n;              % step size
```

After we initialize the values of t, x, and y,

```
» t = t0;      x = x0;      y = y0;
» result = [t0,x0,y0];
```

the Runge-Kutta method itself is implemented by the **for** loop below, which carries out the iteration

$$
\begin{aligned}
k_1 &= f(t_n, x_n, y_n), \\
l_1 &= g(t_n, x_n, y_n) \\
k_2 &= f(t_n + \tfrac{1}{2}h, x_n + \tfrac{1}{2}h\,k_1, y_n + \tfrac{1}{2}h\,l_1), \\
l_2 &= g(t_n + \tfrac{1}{2}h, x_n + \tfrac{1}{2}h\,k_1, y_n + \tfrac{1}{2}h\,l_1) \\
k_3 &= f(t_n + \tfrac{1}{2}h, x_n + \tfrac{1}{2}h\,k_2, y_n + \tfrac{1}{2}h\,l_2), \\
l_3 &= g(t_n + \tfrac{1}{2}h, x_n + \tfrac{1}{2}h\,k_2, y_n + \tfrac{1}{2}h\,l_2) \\
k_4 &= f(t_n + h, x_n + h\,k_3, y_n + h\,l_3), \\
l_4 &= f(t_n + h, x_n + h\,k_3, y_n + h\,l_3) \\
k &= \tfrac{1}{6}(k_1 + 2k_2 + 2k_3 + k_4), \\
l &= \tfrac{1}{6}(l_1 + 2l_2 + 2l_3 + l_4) \\
x_{n+1} &= x_n + h\,k \\
y_{n+1} &= y_n + h\,l, \\
t_{n+1} &= t_n + h
\end{aligned}
$$

n times in successtion to take n steps across the interval from $t = t_0$ to $t = t_f$.

```
» for i = 1:n,
     k1 = f(t,x,y);
     l1 = g(t,x,y);                        % left-hand slopes
     k2 = f(t+h/2,x+h*k1/2,y+h*l1/2);
     l2 = g(t+h/2,x+h*k1/2,y+h*l1/2); % 1st midpt slopes
     k3 = f(t+h/2,x+h*k2/2,y+h*l2/2);
     l3 = g(t+h/2,x+h*k2/2,y+h*l2/2); % 2nd midpt slopes
     k4 = f(t+h,x+h*k3,y+h*l3);
     l4 = g(t+h,x+h*k3,y+h*l3);       % right-hand slopes
     k = (k1+2*k2+2*k3+k4)/6;         % average x-slope
     l = (l1+2*l2+2*l3+l4)/6;         % average y-slope
     x = x + h*k;                   % Euler step to update x
     y = y + h*l;                   % Euler step to update y
     t = t + h;                     % update t
     if floor(i/m) == i/m,
        result = [result;[180*t,x,y]];
        end                         % adjoin new row of data
     end
```

Then the command

```
» result
result =
         0     1.0000          0
   15.0000     0.9659     0.2588
   30.0000     0.8660     0.5000
   45.0000     0.7071     0.7071
   60.0000     0.5000     0.8660
   75.0000     0.2588     0.9659
   90.0000     0.0000     1.0000
```

displays our table of Runge-Kutta results.

Note that we have arranged to print angles in *degrees* in the first column of output. The second and third columns give the corresponding approximate values of the cosine and sine functions. Noting the familiar values $\cos 60° = \sin 30° = \frac{1}{2}$, $\cos 30° = \sin 60° = \frac{1}{2}\sqrt{3} \approx 0.8660$, and $\cos 45° = \sin 45° = \frac{1}{2}\sqrt{2} \approx 0.7071$, we see that the Runge-Kutta method with only $n = 12$ subintervals has provided 4 decimal places of accuracy on the whole range from 0º to 90º.

For a different initial value problem, we need only re-define the functions $f(x,y)$ and $g(x,y)$ corresponding to our new differential equations and the desired initial and final values in the initial commands above, then re-execute the subsequent ones.

Project 18

Applications of Numerical Methods

Reference: Section 4.3 of Edwards & Penney
DIFFERENTIAL EQUATIONS with Computing and Modeling

The investigations of this section are intended as applications of the more sophisticated numerical DE approximation methods that are "built into" systems such as *Maple, Mathematica*, and MATLAB (as opposed to the *ad hoc* Runge-Kutta methods of the previous project.) We illustrate these high-precision methods by applying them
to analyze the simple initial value problem

$$\frac{dx}{dt} = -8y, \qquad \frac{dy}{dt} = 2x$$
$$x(0) = 10, \qquad y(0) = 0 \tag{1}$$

whose exact solution is given by

$$x(t) = 10\cos 4t, \qquad y(t) = 5\sin 4t. \tag{2}$$

Investigation A below is your own personal ellipse problem, and will help to prepare you for the more substantial investigations that follow.

Investigation A: Your Own Ellipse

Let a be the largest and b the smallest nonzero digit of your student ID number. Then use your system's numerical DE solver to approximate the solution of the initial value problem defined by

$$x' = -a\,y, \qquad y' = b\,x$$

with $x(0)$ being the next smallest digit and $y(0)$ the next largest digit of your ID. In particular, determine numerically the major and minor semi-axes of the ellipse that is parametrized by $(x(t), y(t))$, as well as the period of motion around this ellipse. Are your numerical results consistent with the exact symbolic solution found using your system's "dsolve" function?

Investigation B: The Trajectory of a Baseball

To investigate the trajectory of a baseball batted or thrown into the air with initial velocity vector $\mathbf{v} = (v_{0x}, v_{0y})$, let $x(t)$ denote the horizontal distance the ball has traveled and $y(t)$ its height above the ground after t seconds. Then the velocity components $p = x'(t)$ and $q = y'(t)$ satisfy the initial value problem

$$
\begin{aligned}
x' &= p, & x(0) &= 0 \\
y' &= q, & y(0) &= 0 \\
p' &= -c\,p\sqrt{p^2+q^2}, & p(0) &= v_{0x} \\
q' &= -c\,q\sqrt{p^2+q^2}-g, & q(0) &= v_{0y}.
\end{aligned}
$$

Here $g = 32\ \text{ft/sec}^2$ denotes gravitational acceleration and c is an empirical air resistance coefficient.

(a) First test your methods with $c = 0$, corresponding to the case of *no* air resistance. If the ball leaves the ground at point $(0, 0)$ with initial velocity 160 ft/sec at an inclination of 30° to the ground, then an easy exact solution shows it should travel $400\sqrt{3}$ feet horizontally and hit the ground after $t = 5$ seconds, having reached a maximum height in flight of 100 feet. Check these results numerically. You need to focus on the points of the trajectory where the derivative $y'(t)$ changes sign (the ball's apex) and where $y(t)$ itself changes sign (the ball's impact with the ground).

(b) Now suppose that $c = 0.0025$. Determine numerically the ball's horizontal range, its total time of flight, its maximum height above the ground, and when attained. You should find that air resistance has transformed a massive home run into a routine fly ball.

Investigation C: Kepler's Laws of Planetary (or Satellite) Motion

Consider a satellite in elliptical orbit around a planet of mass M, and suppose that physical units are so chosen that $GM = 1$ (where G is the gravitational constant). If the planet is located at the origin in the xy-plane, then the equations of motion of the satellite are

$$
\frac{d^2x}{dt^2} = -\frac{x}{\left(x^2+y^2\right)^{3/2}}, \qquad \frac{d^2y}{dt^2} = -\frac{y}{\left(x^2+y^2\right)^{3/2}}. \tag{3}
$$

Let T denote the period of revolution of the satellite in its orbit. Kepler's third law says that the *square* of T is proportional to the *cube* of the major semiaxis a of its elliptical orbit. In particular, if $GM = 1$, then

$$
T^2 = 4\pi^2 a^3. \tag{4}
$$

(For details, see Section 12.5 of Edwards and Penney, *Calculus and Analytic Geometry*, 4th ed. (Englewood Cliffs, N.J.: Prentice Hall, 1994).) If the satellite's x- and y-components of velocity, $x_3 = x' = x_1'$ and $x_4 = y' = x_2'$, are introduced, then the system in (3) translates into a system of four first-order differential equations having the form of those in Eq. (22) of Section 4.3 in the text.

(a) Solve this 4-by-4 system numerically with the initial conditions

$$x(0) = 1, \quad y(0) = 0, \quad x'(0) = 0, \quad y'(0) = 1$$

that correspond theoretically to a circular orbit of radius $a = 1$, in which case Eq. (4) gives $T = 2\pi$. Is this what you get?

(b) Now solve the system numerically with the initial conditions

$$x(0) = 1, \quad y(0) = 0, \quad x'(0) = 0, \quad y'(0) = \tfrac{1}{2}\sqrt{6}$$

that correspond theoretically to an elliptical orbit with major semiaxis $a = 2$, so Eq. (4) gives $T = 4\pi\sqrt{2}$. Is this what you get?

(c) Investigate what happens when both the x-component and the y-component of the initial velocity are nonzero.

Investigation D: Newton's Nose Cone

In his great *Principia Mathematica* of 1687, Isaac Newton not only showed in Book I that Kepler's laws imply the inverse-square law of gravitation, but proceeded in Book II to investigate a variety of "practical" problems, including the motion of projectiles subject to air resistance. Figure 4.3.13 in the text illustrates a projectile shaped roughly like a bullet with a "nose cone" having radius r and height h. This nose cone is a surface of revolution determined by the curve $y = y(x)$. Newton showed that, on the basis of plausible assumptions about the nature of air resistance, the force F of air resistance to the projectile moving at velocity v is given by

$$F = 2k\rho A v^2 \tag{5}$$

where ρ denotes the density of the air, $A = \pi r^2$ the cross-sectional area of the projectile, and the crucial constant k is given by

$$k = \int_0^r \frac{2\pi x\, dx}{1 + y'(x)^2}. \tag{6}$$

The table in Fig. 4.3.14 of the text lists numerical values of k for various shapes of nose cones with $r = h$ that Newton investigated. It may seem surprising that a "rounded" hemisphere and a "pointed" cone provide the same value $k = 0.5$;

the appropriate parabola — for which $y(r) = h$ — yields the smaller value $k = \frac{1}{4}\ln 5 \approx 0.4024$. But even more surprising is the fact that the optimally angled "flat-tipped" conical frustum (Fig. 4.3.15) yields the still smaller value $k \approx 0.3820$. Newton proceeded to study the much more difficult question of the optimally curved nose-cone-with-flat-tip (Fig. 4.3.16). He founded that the minimum possible air resistance coefficient $k \approx 0.3748$ is obtained when the shape function $y(x)$ satisfies the endpoint value problem

$$y''(x) = \frac{y'(x)\left[1 + y'(x)^2\right]}{x\left[3y'(x)^2 - 1\right]}, \tag{7}$$

$$y(x_0) = 0, \qquad y'(x_0) = 1 \tag{8a}$$

$$y(r) = h. \tag{8b}$$

(a) For a numerical investigation, take $r = h = 1$. The crux of the matter is the determination of the tip radius x_0 so that the solution of the initial value problem defined by Eqs. (7) and (8a) satisfies also the endpoint condition $y(1) = 1$ in (8b). Do this by the method of "shooting", as follows. First convert (7) to a system of two first-order differential equations in $y_1 = y$ and $y_2 = y'$. Start with with $x_0 = 0.5$, say, and solve the resulting initial value problem with the initial conditions $y_1(x_0) = 0$, $y_2(x_0) = 1$. Note that Eqs. (7) and (8a) imply that $y_1''(x) = y_2'(x) > 0$ for $x > x_0$. It follows that the graph of $y_1(x)$ is concave upward on the interval $[x_0, 1]$, as indicated in Fig. 4.3.16 in the text. Do you see why this implies that, if the solution value $y_1(1)$ is less than the target value 1, then the starting value x_0 was too large (why?) and therefore should be decreased for the next "shot"?. Once you have the desired starting value bracketed, proceed by a bisection method — splitting the difference at each stage — until you have determined the appropriate value of x_0 accurate to 4 decimal places. Then find the corresponding optimal value of k by substituting the resulting numerical solution $y_2(x)$ for $y'(x)$ in (6), finally evaluating the integral numerically.

(b) Suppose that a projectile with hemispherical nose (so $k = 0.5$) is fired at an initial inclination angle of 45° with initial velocity 300 ft/sec and travels 2000 feet before striking the ground. What is the value for this projectile of the constant c in Eq. (23) of the text? (Take $g = 32$ ft/s^2.) Answer this question by the "shooting" or "mortar fire" method — solve the appropriate initial value problem numerically with different values of c until you close in on a satisfactory value giving the desired range of 2000 ft.

(c) Now the same projectile is fitted with an optimal conical frustum nose cone, so $k = 0.3820$ and therefore c is $(0.3820)/(0.5)$ times the value found in part (b) (why?). Try to guess in advance the order of magnitude (10 ft or 100 ft?) of the increase in horizontal range due to flattening the tip of the projectile.

(d) Repeat part (c), but now use Newton's optimal nose cone with $k \approx 0.3748$.

Using *Maple*

To investigate numerically the initial value problem in (1), we first define the differential equations

```
> deq1 := diff(x(t),t) = -8*y(t);
> deq2 := diff(y(t),t) =  2*x(t);
```

A quick plot with initial values $[t_0, x_0, y_0] = [0,10,0]$ is generated by the command

```
> with(DEtools):
  DEplot([deq1,deq2], [x,y],
         t=0..5, {[0,10,0]},
         x=-10..10, y=-10..10,
         stepsize=0.02);
```

We appear (in the first figure on the next page) to have the expected elliptical trajectory with major and minor semi-axes 10 and 5. To estimate the period of revolution of a solution point around this ellipse, we generate a *ty*-graph with the command

```
> DEplot([deq1,deq2], [x,y],
    t=0..5, {[0,10,0]}, y=-5..5,
    stepsize=0.02, scene=[t,y]);
```

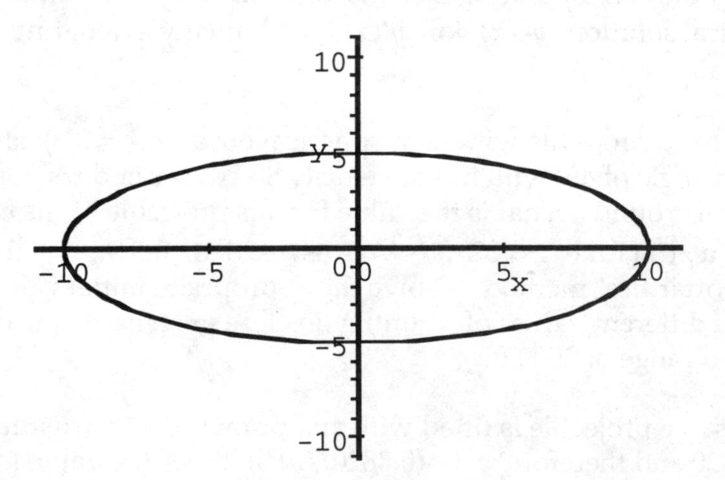

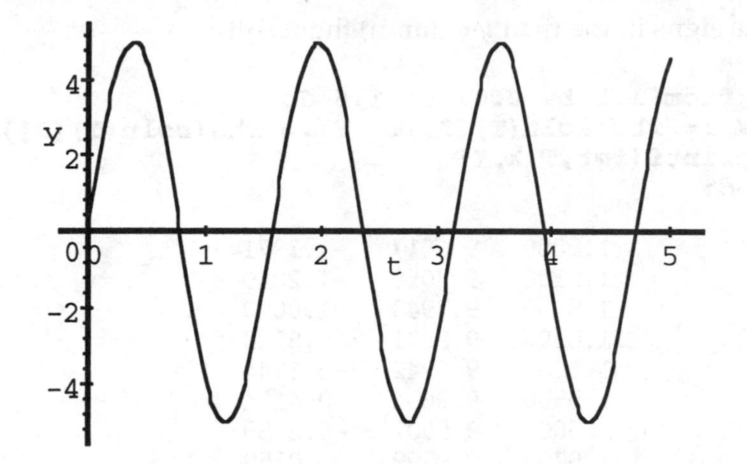

It appears that, having started at $y(0) = 0$, $y(t)$ returns to 0 for the second time somewhere between $t = 1$ and $t = 2$, at approximately $t \approx 1.5$. To focus in on this point, we need a high-precision numerical solution. With

```
> deqs := {deq1,deq2}:    inits := {x(0)=10, y(0)=0}:
```

the command

```
> soln := dsolve(deqs union inits, {x(t),y(t)},
                 type=numeric);

soln := proc(rkf45_x) ... end
```

generates a procedure that computes numerical values of an approximate solution. For instance, the values

```
> soln(1.5);

[t = 1.5, x(t) = 9.601703020132273, y(t) = -1.397077518350383]

> soln(1.6);

[t = 1.6, x(t) = 9.931849355853455, y(t) = .5827460291390618]
```

(looking for the change in sign of y) confirm our graphical observation that the period of motion around the elliptical orbit is between $t = 1.5$ and $t = 1.6$.

To narrow this gap, we can write a simple **do** loop to print table of *txy*-values using the formatting string

```
> fmt := `%6.4f    %6.4f    %6.4f \n`:
```

The (scanning the signs in the final y-column) the result

```
> for T from 1.5 by 0.01 to 1.6 do
      X := rhs(soln(T)[2]):   Y := rhs(soln(T)[3]):
      printf(fmt,T,X,Y)
      od:

                 1.5000   9.6017   -1.3971
                 1.5100   9.7058   -1.2040
                 1.5200   9.7943   -1.0090
                 1.5300   9.8671   -0.8123
                 1.5400   9.9242   -0.6144
                 1.5500   9.9654   -0.4154
                 1.5600   9.9907   -0.2159
                 1.5700   9.9999   -0.0159
                 1.5800   9.9932    0.1840
                 1.5900   9.9705    0.3837
                 1.6000   9.9318    0.5827
```

tells us that the actual period lies between $t = 1.57$ and $t = 1.58$. Finally, the two further tabulations

```
> for T from 1.57 by 0.001 to 1.58 do
      X := rhs(soln(T)[2]):   Y := rhs(soln(T)[3]):
      printf(fmt,T,X,Y)
      od:

                 1.5700    9.9999  -0.0159
                 1.5710   10.0000   0.0041
                 1.5720    9.9999   0.0241
                 1.5730    9.9996   0.0441
                 1.5740    9.9992   0.0641
                 1.5750    9.9986   0.0841
                 1.5760    9.9978   0.1041
                 1.5770    9.9969   0.1241
                 1.5780    9.9958   0.1441
                 1.5790    9.9946   0.1640
                 1.5800    9.9932   0.1840

> for T from 1.570 by 0.0001 to 1.571 do
      X := rhs(soln(T)[2]):   Y := rhs(soln(T)[3]):
      printf(fmt,T,X,Y)
      od:

                 1.5700    9.9999   -0.0159
                 1.5701   10.0000   -0.0139
                 1.5702   10.0000   -0.0119
                 1.5703   10.0000   -0.0099
                 1.5704   10.0000   -0.0079
                 1.5705   10.0000   -0.0059
                 1.5706   10.0000   -0.0039
```

```
1.5707    10.0000    -0.0019
1.5708    10.0000     0.0001
1.5709    10.0000     0.0021
1.5710    10.0000     0.0041
```

show that the period is 1.5708 accurate to 4 decimal places (consistent with the fact that the exact solution has period $\pi/2$).

Using *Mathematica*

To investigate numerically the initial value problem in (1), we first define the differential equations

```
deq1 = D[x[t],t] == -8*y[t];
deq2 = D[y[t],t] ==  2*x[t];
```

Then the command

```
soln = NDSolve[ {deq1, deq2,
                 x[0] == 10, y[0] == 0},
                {x, y}, {t,0,10} ]

{{x -> InterpolatingFunction[{0., 10.}, <>],
  y -> InterpolatingFunction[{0., 10.}, <>]}}
```

generates "interpolating functions" that approximate the coordinate functions of an xy-solution with the initial values $x(0) = 10$ and $y(0) = 0$. The command

```
ParametricPlot[ Evaluate[ {x[t],y[t]} /. soln ],
                {t,0,10} ]
```

then generates a quick phase plane plot of this approximate solution.

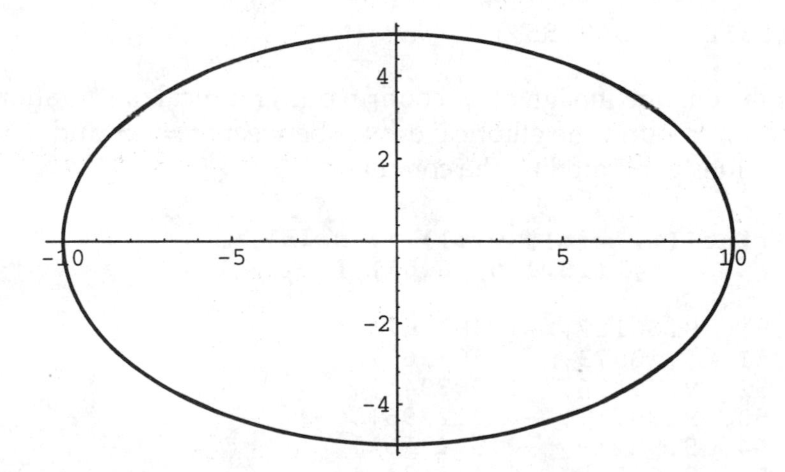

We appear to have the expected elliptical trajectory with major and minor semi-axes 10 and 5. To estimate the period of revolution of a solution point around this ellipse, we generate a ty-graph with the command

```
Plot[ Evaluate[ y[t] /. soln ], {t,0,5} ]
```

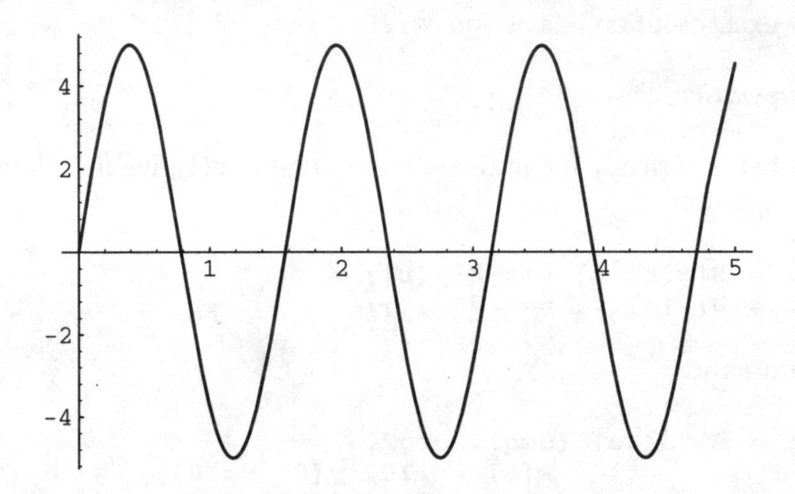

It appears that, having started at $y(0) = 0$, $y(t)$ returns to 0 for the second time somewhere between $t = 1$ and $t = 2$, at approximately $t \approx 1.5$.

To focus in on this point, we calculate numerical values of our approximate solution. For instance, the values

```
{x[1.5], y[1.5]} /. soln
{{9.60167, -1.39706}}

{x[1.6], y[1.6]} /. soln
{{9.9318, 0.582755}}
```

(looking for the change in sign of y) confirm our graphical observation that the period of motion around the elliptical orbit is between $t = 1.5$ and $t = 1.6$. Then the table of values generated by the command

```
Do[Print[{t, x[t], y[t]} /. soln],
        {t, 1.5,1.6, 0.01} ]

{{1.5,  9.60167, -1.39706}}
{{1.51, 9.70573, -1.20396}}
{{1.52, 9.79426, -1.00894}}
{{1.53, 9.86711, -0.812296}}
{{1.54, 9.92419, -0.614356}}
{{1.55, 9.96538, -0.415434}}
```

```
{{1.56, 9.99064, -0.215847}}
{{1.57, 9.99991, -0.0159147}}
{{1.58, 9.99318,  0.184043}}
{{1.59, 9.97047,  0.383706}}
{{1.6,  9.9318,   0.582755}}
```

tells us that the actual period lies between $t = 1.57$ and $t = 1.58$. Finally, the two further tabulations

```
Do[Print[{t, x[t], y[t]} /. soln],
    {t, 1.57,1.58, 0.001} ]

{{1.57,  9.99991, -0.0159147}}
{{1.571, 9.99996,  0.00408522}}
{{1.572, 9.99984,  0.024085}}
{{1.573, 9.99957,  0.0440845}}
{{1.574, 9.99914,  0.0640832}}
{{1.575, 9.99854,  0.0840809}}
{{1.576, 9.99779,  0.104077}}
{{1.577, 9.99688,  0.124072}}
{{1.578, 9.99581,  0.144065}}
{{1.579, 9.99457,  0.164055}}
{{1.58,  9.99318,  0.184043}}
```

```
Do[Print[{t, x[t], y[t]} /. soln],
    {t, 1.570,1.571, 0.0001} ]

{{1.57,   9.99991, -0.0159147}}
{{1.5701, 9.99992, -0.0139147}}
{{1.5702, 9.99993, -0.0119147}}
{{1.5703, 9.99994, -0.00991472}}
{{1.5704, 9.99995, -0.00791473}}
{{1.5705, 9.99995, -0.00591474}}
{{1.5706, 9.99996, -0.00391475}}
{{1.5707, 9.99996, -0.00191476}}
{{1.5708, 9.99996,  0.000085235}}
{{1.5709, 9.99996,  0.00208523}}
{{1.571,  9.99996,  0.00408522}}
```

show that the period is 1.5708 accurate to 4 decimal places (consistent with the fact that the exact solution has period $\pi/2$).

Using MATLAB

MATLAB includes the low order solver **ode23** and the higher order solver **ode45**. The low order method suffices for most graphical purposes. To investigate numerically the initial value problem in (1), we first define MATLAB function

```
function yp = ypellipse(t,y)
yp = y;
yp(1) = -8*y(2);
yp(2) =  2*y(1);
```

that describes the system in the vector form $y' = Ay$ with $y = \begin{bmatrix} y_1 \\ y_2 \end{bmatrix}$.

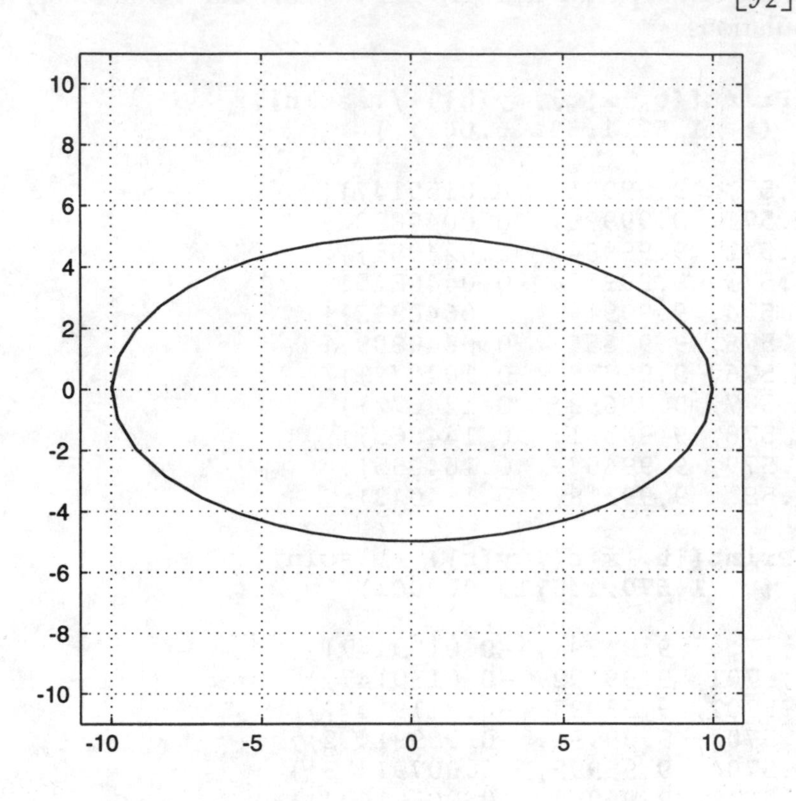

Then the command

```
» [T,Y] = ode23('ypellipse', 0,2*pi, [10;0]);
```

generates a column vector **T** of t-values and a matrix **Y** whose two column vectors approximate the corresponding values of the coordinate functions of a solution with the initial values $y_1(0) = 10$ and $y_2(0) = 0$. The commands

```
» plot(Y(:,1),Y(:,2))
» axis([-11 11 -11 11])
» axis square, hold on, grid on
```

then generate a quick phase plane plot of this approximate solution. We appear to have the expected elliptical trajectory with major and minor semi-axes 10 and 5. To estimate the period of revolution of a solution point around this ellipse, we generate a ty-graph with the commands

```
» plot(T,Y(:,2))
» axis([0 5 -5 5])
» axis square, hold on, grid on
```

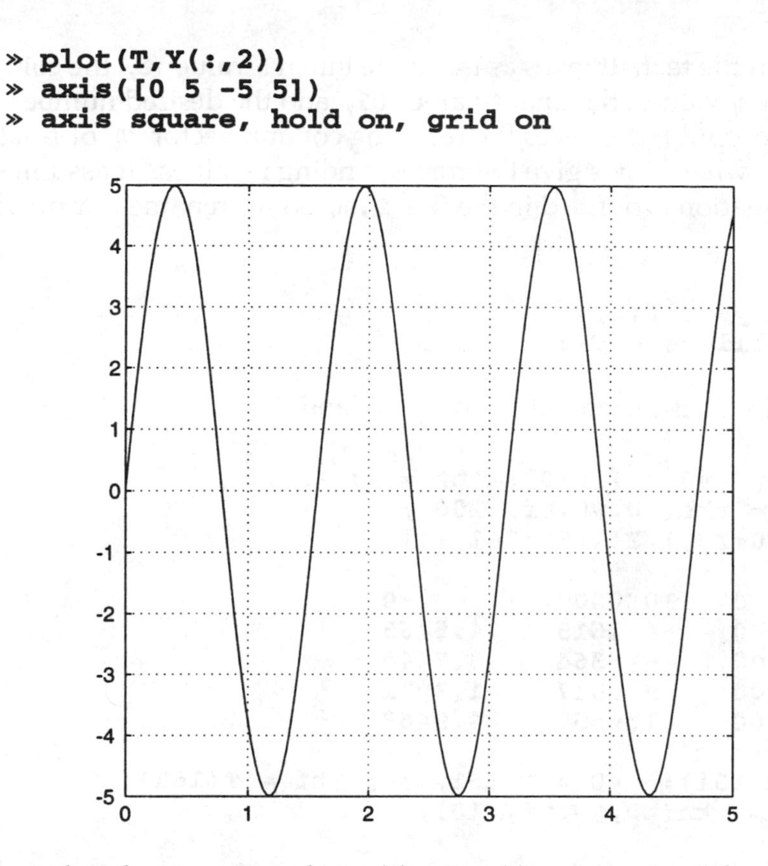

It appears that, having started at $y(0) = 0$, $y(t)$ returns to 0 for the second time somewhere between $t = 1$ and $t = 2$, at approximately $t \approx 1.5$.

To focus in on this point, we calculate numerical values of our approximate solution. For this purpose we can use the *ad hoc* *n*-dimensional Runge-Kutta function

```
function  [T,Y] = rkn(t,y,t1,n)
h = (t1 - t)/n;                          % step size
T = t;                                   % initial t
Y = y';                                  % initial y-vector
for i = 1 : n                            % begin loop

     k1 = f(t, y);                       % first k-vector
     k2 = f(t+h/2,y+h*k1/2);             % second k-vector
     k3 = f(t+h/2,y+h*k2/2);             % third k-vector
     k4 = f(t+h  ,y+h*k3  );             % fourth k-vector
     k  = (k1+2*k2+2*k3+k4)/6;           % average k-vector
     t  = t + h;                         % new t
     y  = y + h*k;                       % new y
     T = [T;t];                          % update t-column
     Y = [Y;y'];                         % update y-matrix
     end                                 % end loop
```

that is described in the text. It takes as input the initial t-value **t**, the column vector **y** of initial y-values, the final t-value **t1**, and the desired number **n** of subintervals. As output it produces the resulting column vector **T** of t-values and the matrix **Y** whose rows give the corresponding y-values. It assumes that the differential equation is defined in the file **f.m**, so we rename our previous function:

```
function yp = f(t,y)
yp = ypellipse(t,y);
```

Then the tables of values generated by the commands

```
» t0 = 0;   y0 = [10;0];   tf = 2;
» [T,Y] = rkn(t0,y0,tf, 200);
» [T(1:50:201),Y(1:50:201,:)]
ans =
          0   10.0000         0
     0.5000   -4.1615    4.5465
     1.0000   -6.5364   -3.7840
     1.5000    9.6017   -1.3971
     2.0000   -1.4550    4.9468

» t0 = T(151);   y0 = Y(151,:)';   tf = T(161);
» [T,Y] = rkn(t0,y0,tf, 10);
» [T,Y]
ans =
     1.5000    9.6017   -1.3971
     1.5100    9.7058   -1.2040
     1.5200    9.7943   -1.0090
     1.5300    9.8671   -0.8123
     1.5400    9.9242   -0.6144
     1.5500    9.9654   -0.4154
     1.5600    9.9907   -0.2159
     1.5700    9.9999   -0.0159
     1.5800    9.9932    0.1840
     1.5900    9.9705    0.3837
     1.6000    9.9318    0.5827
```

(scanning the final column looking for the change in sign of y_2) confirm our graphical observation that the period of motion around the elliptical orbit lies between $t = 1.5$ and $t = 1.6$, indeed, between $t = 1.57$ and $t = 1.59$. Next, the successive tables

```
» t0 = T(8);   y0 = Y(8,:)';   tf = T(9);
» [T,Y] = rkn(t0,y0,tf, 10);
» [T,Y]
ans =
     1.5700    9.9999   -0.0159
     1.5710   10.0000    0.0041
     1.5720    9.9999    0.0241
     1.5730    9.9996    0.0441
```

```
1.5740     9.9992     0.0641
1.5750     9.9986     0.0841
1.5760     9.9978     0.1041
1.5770     9.9969     0.1241
1.5780     9.9958     0.1441
1.5790     9.9946     0.1640
1.5800     9.9932     0.1840
```

and

```
» t0 = T(1);  y0 = Y(1,:)';  tf = T(2);
» [T,Y] = rkn(t0,y0,tf, 10);
» [T,Y]
ans =
    1.5700     9.9999    -0.0159
    1.5701    10.0000    -0.0139
    1.5702    10.0000    -0.0119
    1.5703    10.0000    -0.0099
    1.5704    10.0000    -0.0079
    1.5705    10.0000    -0.0059
    1.5706    10.0000    -0.0039
    1.5707    10.0000    -0.0019
    1.5708    10.0000     0.0001
    1.5709    10.0000     0.0021
    1.5710    10.0000     0.0041
```

show that the period is 1.5708 accurate to 4 decimal places (consistent with the fact that the exact solution has period $\pi/2$).

Earth-Moon Satellite Orbits

We consider finally an Apollo satellite in orbit about the Earth E and the Moon M. Figure 4.3.8 in the text shows an x_1x_2-coordinate system whose origin lies at the center of mass of the Earth and the Moon, and which rotates at the rate of one revolution per "moon month" of approximately $\tau = 27.32$ days, so the Earth and Moon remain fixed in their positions on the x_1-axis. If we take as unit distance the distance between the Earth and Moon centers, then their coordinates are $E(-\mu, 0)$ and $M(1-\mu, 0)$, where $\mu = m_M/(m_E + m_M)$ in terms of the Earth mass m_E and the Moon mass m_M. If we take the total mass $m_E + m_M$ as the unit of mass and $\tau/2\pi \approx 4.348$ days as the unit of time, then the gravitational constant has value $G = 1$, and the equations of motion of the satellite position $S(x_1, x_2)$ are

$$x_1'' = x_1 + 2x_2' - \frac{(1-\mu)(x_1+\mu)}{r_E^3} - \frac{\mu(x_1-1+\mu)}{r_M^3} \tag{1a}$$

and

$$x_2'' = x_2 - 2x_1' - \frac{(1-\mu)x_2}{r_E^3} - \frac{\mu x_2}{r_M^3} \tag{1b}$$

where $r_E = \sqrt{(x_1+\mu)^2 + x_2^2}$ and $r_M = \sqrt{(x_1+\mu-1)^2 + x_2^2}$ denote the satellite's distance to the Earth and Moon, respectively. The initial two terms on the right-hand side of each equation in (1) result from the rotation of the coordinate system. In the system of units described here, the lunar mass is approximately $\mu = m_M = 0.012277471$. The second-order system in (1) can be converted to a first-order system by substituting

$$x_1' = x_3, \qquad x_2' = x_4$$

so

$$x_3' = x_1'', \qquad x_4' = x_2''. \tag{2}$$

This system is defined in the MATLAB function

```
function  yp = ypmoon(t,y)

%  Equations of motion of small satellite in rotating
%  earth-moon center-of-mass system.  The earth is
%  located at  (-m1,0)  and the moon at  (1-m1,0).

%  y(1) and y(2) = coordinates of the satellite.
%  y(3) and y(4) = velocity components of satellite.

m1 = 0.012277471;                         % mass of moon
m2 = 0.987722529;                         % mass of earth

r1 = norm([y(1)+m1, y(2)]);               % Distance to the
earth
r2 = norm([y(1)-m2, y(2)]);               % Distance to the moon

yp = [ y(3); y(4); 0; 0 ];                % Column 4-vector

yp(3) = y(1)+2*y(4) - m2*(y(1)+m1)/r1^3 - m1*(y(1)-
m2)/r2^3;

yp(4) = y(2)-2*y(3) - m2*y(2)/r1^3 - m1*y(2)/r2^3;
```

Suppose that the satellite initially is in a clockwise circular orbit of radius about 1500 miles about the Moon. At its farthest point from the Earth ($x_1 = 0.994$) it is "launched" into Earth-Moon orbit with initial velocity v_0. We then want to solve the system in (2) — with the right-hand functions in (1) substituted for x_1'' and x_2'' — with the initial conditions

$$x_1(0) = 0.994, \qquad x_2(0) = 0, \qquad x_3(0) = 0, \qquad x_4(0) = -v_0. \tag{3}$$

In the system of units used here, the unit of velocity is approximately 2289 miles per hour. Some initial velocities and final times of particular interest are defined by the file

```
function  [tf,y0] = mooninit(k)

% Initial conditions for k-looped Apollo orbit

if  k == 2,
    tf = 5.436795439260;
    y0 = [ 0.994 0 0 -2.113898796695 ]';
elseif  k == 3,
    tf = 11.124340337266;
    y0 = [ 0.994 0 0 -2.031732629557 ]';

elseif  k == 4,
    tf = 17.065216560158;
    y0 = [ 0.994 0 0 -2.001585106379 ]';
    end
```

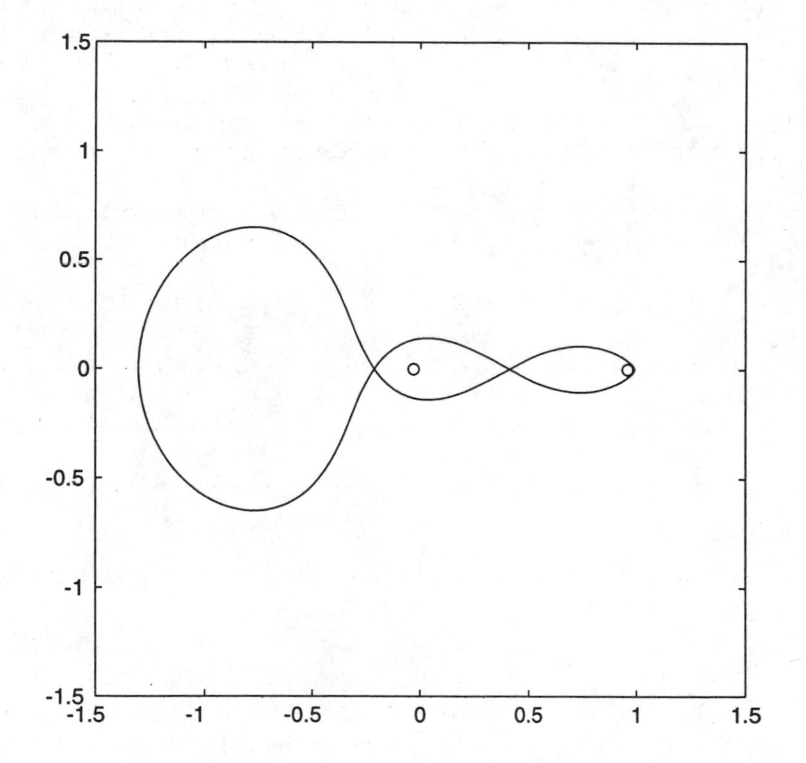

The cases $k = 3$ and $k = 3$ yield Figures 4.3.9 and 4.3.10 (respectively) in the text. The following commands (with $k = 2$) yield the figure above, and illustrate how such figures are plotted.

```
» [tf,y0] = mooninit(2);
» [T,Y] = ode45('ypmoon',0,tf, y0);
» ts = 0 : tf/500 : tf;
» xs = spline(T, Y(:,1), ts);
» ys = spline(T, Y(:,2), ts);
» plot(xs,ys)
```

The high order solver **`ode45`** yields points that are too far apart for a smooth plot, so we have "splined" the results to interpolate more points (see **`help spline`** for an explanation). You might like to try the values $k = 3$ and $k = 4$ to generate the analogous 3- and 4-looped orbits. A more substantial project would be to search empirically for initial velocities yielding periodic orbits with more than 4 loops.

Chapter 5

Linear Systems of Differential Equations

Project 19
Matrix Calculations with Computer Systems

Reference: Section 5.1 of Edwards & Penney
DIFFERENTIAL EQUATIONS with Computing and Modeling

Calculations with numerical matrices of order greater than 3 are most frequently carried out with the aid of calculators or computers. For example, in Example 8 of the text we needed to solve the linear system

$$\begin{aligned} 2c_1 + 2c_2 + 2c_3 &= 0 \\ 2c_1 \qquad - 2c_3 &= 2 \\ c_1 - \ c_2 + \ c_3 &= 6. \end{aligned} \tag{1}$$

Writing this system in matrix notation $A\mathbf{c} = \mathbf{b}$ and working with a TI-85 calculator, for instance, we would store the 3×3 coefficient matrix A and the column vector $\mathbf{b}$ with the commands

```
[[2,2,2] [2,0,-2] [1,-1,1]] → A
[[0] [2] [6]] → b
```

in which each matrix and each of its successive rows in enclosed within square brackets. Then we calculate and store the solution $\mathbf{c} = A^{-1}\mathbf{b}$ with the command

```
A⁻¹*b → c
```

which yields the result

```
[[2 ]
 [-3]
 [1 ]]
```

Thus $c_1 = 2$, $c_2 = -3$, and $c_3 = 1$, as we found using elementary row operations in Example 8 in the text.

Matrix notation and calculation in most computer systems are similar. In the paragraphs below, we illustrate how to enter matrices and perform simple matrix calculations using *Maple*, *Mathematica*, and MATLAB. The "investigations" that follow are simple exercises to familiarize you with automatic matrix computation. Applications will appear in the remaining projects in this chapter.

Investigation A

To practice simple matrix algebra with whatever calculator or computer system is available, you might begin by defining a square matrix A of your own selection. Then calculate its inverse matrix B and check that the matrix product AB is the identity matrix. Do this with several square matrices of different dimensions.

Investigation B

Use matrix algebra as indicated above (for the computations of Example 8) to solve Problems 31 through 40 of this section.

Investigation C

An interesting $n \times n$ matrix is the **Hilbert matrix** H_n whose ijth element (in the ith row and jth column) is $1/(i+j-1)$. For instance, the 4×4 Hilbert matrix is

$$H_4 = \begin{bmatrix} 1 & \frac{1}{2} & \frac{1}{3} & \frac{1}{4} \\ \frac{1}{2} & \frac{1}{3} & \frac{1}{4} & \frac{1}{5} \\ \frac{1}{3} & \frac{1}{4} & \frac{1}{5} & \frac{1}{6} \\ \frac{1}{4} & \frac{1}{5} & \frac{1}{6} & \frac{1}{7} \end{bmatrix}. \qquad (2)$$

Set up the Hilbert matrices of orders $n = 3, 4, 5, 6, \ldots,$ and calculate their inverse matrices. The results will look more interesting than you might anticipate.

(1) Show that the system $H_3\mathbf{x} = \mathbf{b}$ has the exact solution $\mathbf{x} = [1 \;\; 1 \;\; 1]^T$ if

$$\mathbf{b} = [{}^{11}\!/\!_6 \;\; {}^{13}\!/\!_{12} \;\; {}^{47}\!/\!_{60}]^T \approx [1.83333 \;\; 1.08333 \;\; 0.78333]^T,$$

whereas it has the exact solution $\mathbf{x} = [0.6 \;\; 2.4 \;\; 0]^T$ if $\mathbf{b} = [1.8 \;\; 1.1 \;\; 0.8]^T$. Thus linear systems with coefficient matrix H_3 can be quite "unstable" with respect to roundoff errors in the constant vector $\mathbf{b}$. (This example is essentially the one given by Steven H. Weintraub on page 324 of the April 1986 issue of the *American Mathematical Monthly*.)

(2) Show that the system $H_4\mathbf{x} = \mathbf{b}$ has the exact solution $\mathbf{x} = [1 \;\; 1 \;\; 1 \;\; 1]^T$ if

$$\mathbf{b} = \left[\tfrac{25}{12} \quad \tfrac{77}{60} \quad \tfrac{57}{60} \quad \tfrac{319}{420}\right]^T \approx [2.08333 \quad 1.28333 \quad 0.95000 \quad 0.75952]^T,$$

whereas it has the exact solution $\mathbf{x} = [1.28 \quad -1.8 \quad 7.2 \quad -2.8]^T$ if

$$\mathbf{b} = [2.08 \quad 1.28 \quad 0.95 \quad 0.76]^T.$$

Using *Maple*

In order to carry out matrix computations the *Maple* `linalg` package must first be loaded,

```
with(linalg):
```

A particular matrix can be entered either with a command of the form

```
> A := matrix( 3,3, [2,2,2, 2,0,-2, 1,-1,1]);
```

$$A := \begin{bmatrix} 2 & 2 & 2 \\ 2 & 0 & -2 \\ 1 & -1 & 1 \end{bmatrix}$$

where the first two arguments prescribe the numbers of rows and columns and the third argument is a vector listing the elements of A row by row, or with one of the form

```
> A := matrix( [[2,2,2], [2,0,-2], [1,-1,1]] );
```

in which the individual row vectors of A are prescribed. The inverse matrix A^{-1} is then calculated with the command

```
> B := inverse(A);
```

$$B := \begin{bmatrix} \frac{1}{8} & \frac{1}{4} & \frac{1}{4} \\ \frac{1}{4} & 0 & \frac{-1}{2} \\ \frac{1}{8} & \frac{-1}{4} & \frac{1}{4} \end{bmatrix}$$

The function `evalm` (for **eval**uate **m**atrices) is used for matrix multiplication with `&*` denoting the matrix multiplication operator itself. Thus we can verify that B actually is the inverse matrix of A with the calculation

```
> evalm( B &* A );
```

$$\begin{bmatrix} 1 & 0 & 0 \\ 0 & 1 & 0 \\ 0 & 0 & 1 \end{bmatrix}$$

Having defined the right-hand side constant (column) vector

```
> b := matrix( 3,1, [0,2,6]);
```

$$b := \begin{bmatrix} 0 \\ 2 \\ 6 \end{bmatrix}$$

in (1), we can then calculate the solution vector $\mathbf{c} = A^{-1}\mathbf{b} = B\mathbf{b}$ with the command

```
> c := evalm( B &* b );
```

$$c := \begin{bmatrix} 2 \\ -3 \\ 1 \end{bmatrix}$$

For higher-dimensional linear systems the computation of the inverse matrix is not so efficient as immediate application of *Maple*'s **linsolve** function:

```
> c := linsolve( A, b );
```

$$c := \begin{bmatrix} 2 \\ -3 \\ 1 \end{bmatrix}$$

A matrix can also be defined by prescribing its *ij*th element as a function of the row index i and the column index i. For instance, the 3×3 Hilbert matrix is defined by

```
> H3 := matrix( 3,3, (i,j) -> 1/(i+j-1) );
```

$$H3 := \begin{bmatrix} 1 & \frac{1}{2} & \frac{1}{3} \\ \frac{1}{2} & \frac{1}{3} & \frac{1}{4} \\ \frac{1}{3} & \frac{1}{4} & \frac{1}{5} \end{bmatrix}$$

The linalg package also includes an explicit Hilbert matrix function. For instance, the command

```
> H4 := hilbert(4);
```

generates the 4×4 Hilbert matrix displayed in (2). Finally, the **diag**onal function can be used to generate the most ubiquitous of all special matrices — the identity matrices such as

```
> I3 = diag(1,1,1);
```

$$\begin{bmatrix} 1 & 0 & 0 \\ 0 & 1 & 0 \\ 0 & 0 & 1 \end{bmatrix}$$

Using *Mathematica*

A particular matrix is entered in *Mathematica* as as list of row vectors, each row vector itself being a list of elements, as in the command

```
A = { {2,2,2}, {2,0,-2}, {1,-1,1} }

{{2, 2, 2}, {2, 0, -2}, {1, -1, 1}}
```

The **MatrixForm** function can be used to display A in standard matrix form:

```
A // MatrixForm

2    2    2
2    0    -2
1    -1   1
```

The inverse matrix A^{-1} is then calculated with the command

```
B = Inverse[A]

  1  1  1    1        1     1    1   1
{{-, -, -}, {-, 0, -(-)}, {-, -(-), -}}
  8  4  4    4        2     8    4   4
```

The period **.** denotes the matrix multiplication operator in *Mathematica*. Thus we can verify that B actually is the inverse matrix of A with the calculation

```
B . A // MatrixForm

1   0   0

0   1   0

0   0   1
```

Having defined the right-hand side constant (column) vector

```
b = { {0}, {2}, {6} }

{{0}, {2}, {6}}
```

in (1), we can then calculate the solution vector $\mathbf{c} = A^{-1}\mathbf{b} = B\mathbf{b}$ with the command

```
c = B . b

{{2}, {-3}, {1}}
```

For higher-dimensional linear systems the computation of the inverse matrix is not so efficient as immediate application of *Mathematica*'s **LinearSolve** function:

```
c = LinearSolve[ A, b ]

{{2}, {-3}, {1}}
```

The computation

```
A.c - b

{{0}, {0}, {0}}
```

then verifies that **c** is a solution of the equation $A\mathbf{x} = \mathbf{b}$.

A matrix can also be defined by prescribing its *ij*th element as a function of the row index *i* and the column index *i*. For instance, the 3×3 Hilbert matrix is defined by

```
H3 = Table[ 1/(i+j-1), {i,1,3},{j,1,3} ]

       1  1     1  1  1     1  1  1
{{1, -, -}, {-, -, -}, {-, -, -}}
       2  3     2  3  4     3  4  5

H3 // MatrixForm

     1   1
     -   -
1    2   3

1    1   1
-    -   -
2    3   4

1    1   1
-    -   -
3    4   5
```

Finally, the **DiagonalMatrix** function can be used to generate the most ubiquitous of all special matrices — the identity matrices such as

```
I3 = DiagonalMatrix[ {1,1,1} ]

{{1, 0, 0}, {0, 1, 0}, {0, 0, 1}}
```

Actually, the identity matrix has its own function (of the dimension):

```
IdentityMatrix[3] // MatrixForm

1   0   0

0   1   0

0   0   1
```

Using MATLAB

A particular matrix can be entered either with a command of the form

```
» A = [ [2 2 2]; [2 0 -2]; [1 -1 1] ]
A =
      2      2      2
      2      0     -2
      1     -1      1
```

where the row vectors of A are separated by semicolons, or with one of the simplest possible form

```
» A = [ 2  2  2
        2  0 -2
        1 -1  1 ]
```

in which the matrix is "built" just as it looks The inverse matrix A^{-1} is then calculated with the command

```
» B = inv(A)
B =
     0.1250     0.2500     0.2500
     0.2500          0    -0.5000
     0.1250    -0.2500     0.2500
```

The ordinary MATLAB multiplication operator ***** suffices to multiply matrices of appropriate dimensions. Thus we can verify that B actually is the inverse matrix of A with the calculation

```
» B*A
ans =
        1     0     0
        0     1     0
        0     0     1
```

Having defined the right-hand side constant (column) vector

```
» b = [0; 2; 6]
b =
        0
        2
        6
```

in (1), we can then calculate the solution vector $\mathbf{c} = A^{-1}\mathbf{b} = B\mathbf{b}$ with the command

```
» c = B*b
c =
        2
       -3
        1
```

For higher-dimensional linear systems the computation of the inverse matrix is not so efficient as immediate application of MATLAB's "linear solve" function in the form of the operator **** that denotes "matrix left division":

```
» c = A\b
c =
        2
       -3
        1
```

A matrix can also be defined with a double loop prescribing its ijth element as a function of the row index i and the column index j. For instance, the 3×3 Hilbert matrix H_3 is defined by

```
» for i = 1:3
      for j = 1:3
      H3(i,j) = 1/(i+j-1);
      end
  end
```

using the notation $A(i, j)$ for the ijth element of the matrix A. After specifying "rational formatting" of output, we can display the result as

```
» format rat
» H3
```

```
H3 =
        1          1/2        1/3
      1/2          1/3        1/4
      1/3          1/4        1/5
```

MATLAB also includes an explicit Hilbert matrix function. For instance, the command

```
» hilb(4)
```

generates the 4×4 Hilbert matrix displayed in (2). Finally, the **diag**onal function can be used to generate the most ubiquitous of all special matrices — the identity matrices such as

```
» I3 = diag([1 1 1])
I3 =
        1          0          0
        0          1          0
        0          0          1
```

Actually, the identity matrix has its own MATLAB function (of the dimension):

```
»  eye(3)
ans =
        1          0          0
        0          1          0
        0          0          1
```

Project 20
Brine Tank Applications of Eigenvalue Solutions

Reference: Section 5.2 of Edwards & Penney
DIFFERENTIAL EQUATIONS with Computing and Modeling

Most computational systems offer the capability to find eigenvalues and eigenvectors readily. For instance, for the matrix

$$A = \begin{bmatrix} -0.5 & 0 & 0 \\ 0.5 & -0.25 & 0 \\ 0 & 0.25 & -0.2 \end{bmatrix} \tag{1}$$

of Example 2 in Section 5.2 of the text, the TI-85 commands

```
[[-0.5,0,0][0.5,-0.25,0][0,0,25,-0.2]] → A
eigVl A
                                   {-.2, -.25, -.5}
eigVc A
                                   [[0.00  0.00   1.00]
                                   [0.00  1.00  -2.00]
                                   [1.00 -5.00   1.67]]
```

produce the three eigenvalues -0.2, -0.25, and -0.5 of A and display beneath each its (column) eigenvector. Note that with results presented in decimal form, it is up to us to guess (and verify by matrix multiplication) that the exact eigenvector associated with the eigenvalue $\lambda = -\frac{1}{2}$ is $\mathbf{v} = \begin{bmatrix} 1 & -2 & \frac{5}{3} \end{bmatrix}^T$. The *Maple* commands

```
with(linalg):
A := matrix(3,3, [-0.5,0,0, 0.5,-0.25,0, 0,0.25,-0.2] );
eigenvects(A);
```

the *Mathematica* commands

```
A = {{-0.5,0,0}, {0.5,-0.25,0}, {0,0.25,-0.2}}
Eigensystem[A]
```

and the MATLAB commands

```
[-0.5,0,0; 0.5,-0.25,0; 0,0.25,-0.2]
[V, E] = eig(A)
```

(where E will be a diagonal matrix displaying the eigenvalues of A and the column vectors of V are the corresponding eigenvectors) produce similar results. You can use these commands to find the eigenvalues and eigenvectors needed for any of the problems in Section 5.2 of the text.

Investigation

Consider a linear cascade of 5 full brine tanks whose volumes v_1, v_2, v_3, v_4, v_5 are given by

$$v_i = 10\, d_i \qquad \text{(gallons)}$$

where d_1, d_2, d_3, d_4, d_5 are the first five *distinct* non-zero digits of ***your*** social security number (pick additional ones at random if your social security number has less than five distinct non-zero digits).

Initially, Tank 1 contains one pound of salt per gallon of brine, whereas the remaining tanks contain pure water. The brine in each tank is kept thoroughly mixed, and the flow rate out of each tank is $r_i = 10$ gal/min. Your task is to investigate the subsequent amounts $x_1(t)$, $x_2(t)$, $x_3(t)$, $x_4(t)$, $x_5(t)$ of salt (in pounds) present in these brine tanks.

1. If fresh water flows into Tank 1 at the rate of 10 gal/min, then these functions satisfy the system

$$\begin{aligned} x_1' &= \qquad\quad -k_1x_1 \\ x_2' &= +k_1x_1 - k_2x_2 \\ x_3' &= +k_2x_2 - k_3x_3 \\ x_4' &= +k_3x_3 - k_4x_4 \\ x_5' &= +k_4x_4 - k_5x_5 \end{aligned} \tag{2}$$

where $k_i = r_i/v_i$ for i = 1, 2, ..., 5. Find the eigenvalues $\lambda_1, \lambda_2, \lambda_3, \lambda_4, \lambda_5$ and the corresponding eigenvectors $\mathbf{v}_1, \mathbf{v}_2, \mathbf{v}_3, \mathbf{v}_4, \mathbf{v}_5$ of the this system's coefficient matrix in order to write the general solution in the form

$$\mathbf{x}(t) = c_1\mathbf{v}_1e^{\lambda_1 t} + c_2\mathbf{v}_2e^{\lambda_2 t} + c_3\mathbf{v}_3e^{\lambda_3 t} + c_4\mathbf{v}_4e^{\lambda_4 t} + c_5\mathbf{v}_5e^{\lambda_5 t}. \tag{3}$$

Use the given initial conditions to find the values of the constants c_1, c_2, c_3, c_4, c_5. Then observe that each $x_i(t) \to 0$ as $t \to \infty$, and explain why you would anticipate this result. Plot the graphs of the component functions $x_1(t), x_2(t), x_3(t), x_4(t), x_5(t)$ of $\mathbf{x}(t)$ on a single picture, and finally note (at least as close as the mouse will take you) the maximum amount of salt that is ever present in each tank.

2. If Tank 1 receives as inflow (rather than fresh water) the outflow from Tank 5, then the first equation in (2) is replaced with the equation

$$x_1' = +k_5x_5 - k_1x_1. \tag{4}$$

Assuming the same initial conditions as before, find the explicit solution of the form in (3). Now show that — in this *closed* system of brine tanks — as $t \to \infty$ the salt originally in Tank 1 distributes itself with uniform concentration throughout the various tanks. A plot should make this point rather vividly.

Maple, *Mathematica*, and MATLAB techniques that will be useful for the investigation above are illustrated in the paragraphs that follow. We consider the "open system" of three brine tanks that is shown in Fig. 5.2.1 of the text (see Example 2 of Section 5.2). The vector $\mathbf{x}(t) = [x_1(t) \;\; x_2(t) \;\; x_3(t)]^T$ of salt amounts (in the three tanks) satisfies the linear system

$$\frac{d\mathbf{x}}{dt} = A\mathbf{x} \tag{5}$$

where A is the 3×3 matrix in (1). If initially Tank 1 contains 15 pounds of salt and the other two tanks contain pure water, then the initial vector is $\mathbf{x}(0) = \mathbf{x}_0 = [15 \ \ 0 \ \ 0]^T$.

Using *Maple*

We begin by entering (as indicated previously) the coefficient matrix in (1),

```
> with(linalg):
> A := matrix(3,3,[-0.5,0,0, 0.5,-0.25,0, 0,0.25,-0.2]);
```

$$A := \begin{bmatrix} -0.5 & 0 & 0 \\ 0.5 & -0.25 & 0 \\ 0 & 0.25 & -0.2 \end{bmatrix}$$

and the initial vector

```
> x0 := matrix(3,1, [15,0,0]);
```

$$x0 := \begin{bmatrix} 15 \\ 0 \\ 0 \end{bmatrix}$$

The eigenvalues and eigenvectors of A are calculated with the command

```
eigs := eigenvects(A);
```

$$\begin{aligned} eigs := &[-.5, 1, \{[1 \ \ -2.000000000 \ \ 1.666666667]\}], \\ &[-.25, 1, \{[0 \ \ 1 \ \ -5.000000000]\}], \\ &[-.2, \ 1, \ \{[0 \ \ 0 \ \ 1]\}] \end{aligned}$$

Thus the first eigenvalue λ_1 and its associated eigenvector $\mathbf{v}_1$ are given by

```
> eigs[1][1];
```

$$-.5$$

```
> eigs[1][3][1];
```

$$[1 \quad -2.000000000 \quad 1.666666667]$$

We therefore record the three eigenvalues

```
> L1 := eigs[1][1]:
  L2 := eigs[2][1]:
  L3 := eigs[3][1]:
```

and the corresponding three eigenvectors

```
> v1 := matrix(1,3, eigs[1][3][1]):
  v2 := matrix(1,3, eigs[2][3][1]):
  v3 := matrix(1,3, eigs[3][3][1]):
```

The matrix V with these three column vectors is then defined by

```
> V := transpose(stack(v1,v2,v3));
```

$$V := \begin{bmatrix} 1 & 0 & 0 \\ -2.000000000 & 1 & 0 \\ 1.666666667 & -5.000000000 & 1 \end{bmatrix}$$

To find the constants c_1, c_2, c_3 in the solution

$$\mathbf{x}(t) \;=\; c_1\mathbf{v}_1 e^{\lambda_1 t} + c_2\mathbf{v}_2 e^{\lambda_2 t} + c_3\mathbf{v}_3 e^{\lambda_3 t} \tag{6}$$

we need only solve the system $V\mathbf{c} = \mathbf{x}_0$:

```
> c := linsolve(V,x0);
```

$$c := \begin{bmatrix} 15. \\ 30. \\ 125.0000000 \end{bmatrix}$$

Recording the values of these three constants,

```
> c1 := c[1,1]:   c2 := c[2,1]:   c3 := c[3,1]:
```

we can finally calculate the solution

```
> x := evalm(c1*v1*exp(L1*t) + c2*v2*exp(L2*t) +
             c3*v3*exp(L3*t)):
```

in (6) with component functions

```
> x1 := x[1,1];   x2 := x[1,2];   x3 := x[1,3];
```

$$x1 := 15.\, e^{-.5t}$$

$$x2 := -30.00000000\, e^{-.5t} + 30.\, e^{-.25t}$$

$$x2 := 25.00000001\, e^{-.5t} - 150.00000000\, e^{-.25t} + 125.00000000\, e^{-.2t}$$

The command

```
> plot( {x1,x2,x3}, t = 0..30 );
```

produces the figure below showing the graphs of the functions $x_1(t)$, $x_2(t)$, and $x_3(t)$ giving the amounts of salt in the three tanks. We can approximate the maximum value of each $x_i(t)$ by mouse-clicking on the apex of the appropriate graph.

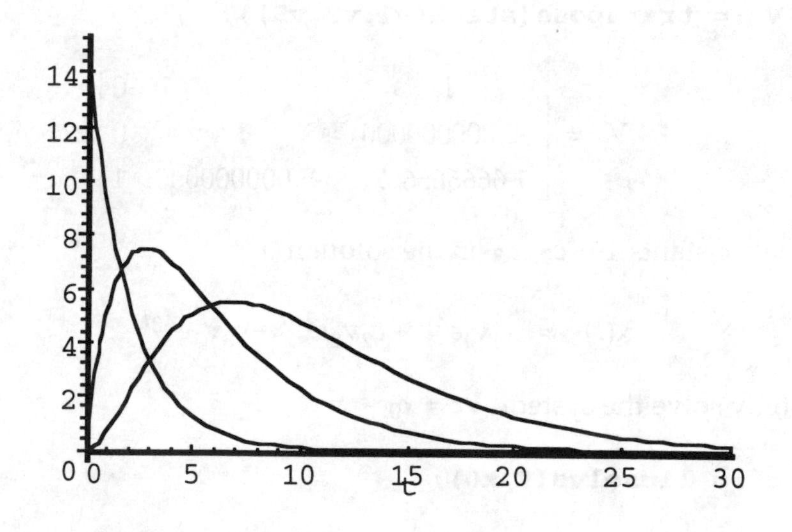

Using *Mathematica*

We begin by entering (as indicated previously) the coefficient matrix in (1),

```
A = {{-0.50 ,0,0}, {0.50,-0.25,0}, {0,0.25,-0.20}};
A // MatrixForm

-0.5    0       0
0.5     -0.25   0
0       0.25    -0.2
```

and the initial vector

```
x0 = {{15}, {0}, {0}};  x0 // MatrixForm

15
0
0
```

The eigenvalues and eigenvectors of A are calculated with the command

```
eigs = Eigensystem[A]

{{-0.5, -0.25, -0.2}, {{0.358569, -0.717137, 0.597614},
   {0., 0.196116, -0.980581}, {0., 0., 1.}}}
```

Thus the first eigenvalue λ_1 and its associated eigenvector $\mathbf{v}_1$ are given by

```
eigs[[1,1]]

-0.5

eigs[[2,1]]

{0.358569, -0.717137, 0.597614}
```

We therefore record the three eigenvalues

```
L1 = eigs[[1,1]];
L2 = eigs[[1,2]];
L3 = eigs[[1,3]];
```

and the corresponding three eigenvectors

```
v1 = eigs[[2,1]];
v2 = eigs[[2,2]];
v3 = eigs[[2,3]];
```

The matrix V having these three eigenvectors as its *column* vectors is then defined by

```
V = Transpose[ {v1, v2, v3} ];
V // MatrixForm

0.358569     0.          0.
-0.717137    0.196116    0.
0.597614     -0.980581   1.
```

To find the constants c_1, c_2, c_3 in the solution

$$\mathbf{x}(t) = c_1\mathbf{v}_1 e^{\lambda_1 t} + c_2\mathbf{v}_2 e^{\lambda_2 t} + c_3\mathbf{v}_3 e^{\lambda_3 t} \tag{6}$$

we need only solve the system $V\mathbf{c} = \mathbf{x}_0$:

```
c = LinearSolve[V,x0]

{{41.833}, {152.971}, {125.}}
```

Recording the values of these three constants,

```
c1 = c[[1,1]];    c2 = c[[2,1]];    c3 = c[[3,1]];
```

we can finally calculate the solution

```
x = c1*v1*Exp[L1*t] + c2*v2*Exp[L2*t] + c3*v3*Exp[L3*t];
```

in (6) with component functions

```
x1 = x[[1]]
x2 = x[[2]]
x3 = x[[3]]

 15.         0.           0.
------ + ------- + ------
 0.5 t    0.25 t     0.2 t
E        E          E

 -30.       30.          0.
------ + ------- + ------
 0.5 t    0.25 t     0.2 t
E        E          E

 25.        150.         125.
------ - ------- + ------
 0.5 t    0.25 t     0.2 t
E        E          E
```

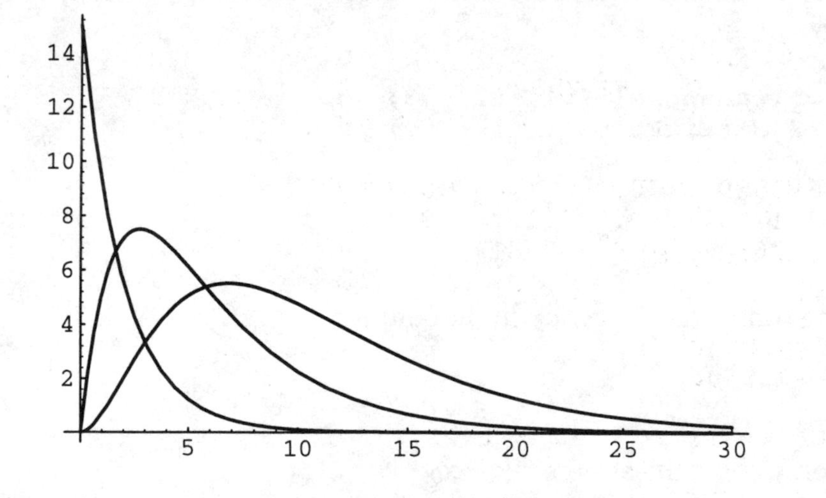

The command

```
Plot[ {x1,x2,x3}, {t,0,30} ]
```

produces the figure above, showing the graphs of the functions $x_1(t)$, $x_2(t)$, and $x_3(t)$ giving the amounts of salt in the three tanks. We can approximate the maximum value of each $x_i(t)$ by mouse-clicking on the apex of the appropriate graph.

Using MATLAB

We begin by entering (as indicated previously) the coefficient matrix in (1),

```
» A = [-0.5      0        0
        0.5   -0.25      0
        0      0.25    -0.2 ]
A =
   -0.5000         0         0
    0.5000   -0.2500         0
         0    0.2500   -0.2000
```

and the initial vector

```
» x0 = [15; 0; 0];
```

The eigenvalues and eigenvectors of A are calculated with the command

```
» [V, E] = eig(A)
V =
         0         0    0.3586
         0    0.1961   -0.7171
    1.0000   -0.9806    0.5976

E =
   -0.2000         0         0
         0   -0.2500         0
         0         0   -0.5000
```

The eigenvalues of A are the diagonal elements

```
» L = diag(E)
L =
   -0.2000
   -0.2500
   -0.5000

» L1 = L(1);      L2 = L(2);      L3 = L(3);
```

of the matrix E. The associated eigenvectors are the corresponding column vectors

```
» v1 = V(:,1);   v2 = V(:,2);   v3 = V(:,3);
```

of the matrix V. To find the constants c_1, c_2, c_3 in the solution

$$\mathbf{x}(t) = c_1\mathbf{v}_1e^{\lambda_1 t} + c_2\mathbf{v}_2e^{\lambda_2 t} + c_3\mathbf{v}_3e^{\lambda_3 t} \tag{6}$$

we need only solve the system $V\mathbf{c} = \mathbf{x}_0$:

```
» c = V\x0;    c'
ans =
  125.0000  152.9706   41.8330
```

Recording the values of these three constants,

```
» c1 = c(1);    c2 = c(2);    c3 = c(3);
```

and defining an appropriate range

```
» t = 0 : 0.1 : 30;
```

of values of t, we can finally calculate the solution

```
» x = c1*v1*exp(L1*t) + c2*v2*exp(L2*t) +
      c3*v3*exp(L3*t);
```

in (6). We plot its three component functions

```
» x1 = x(1,:);    x2 = x(2,:);    x3 = x(3,:);
```

using the command

```
» plot( t,x1, t,x2, t,x3 )
```

The resulting figure (just like those exhibited in the preceding *Maple* and *Mathematica* discussions) shows the graphs of the functions $x_1(t)$, $x_2(t)$, and $x_3(t)$ giving the amounts of salt in the three tanks. We can approximate the maximum value of each $x_i(t)$ by mouse-clicking (after **ginput**) on the apex of the appropriate graph.

Complex Eigenvalues

Finally we consider the "closed system" of three brine tanks that is shown in Fig. 5.2.3 of the text (see Example 4 of Section 5.2). The vector $\mathbf{x}(t) = [x_1(t) \;\; x_2(t) \;\; x_3(t)]^T$ of salt amounts (in the three tanks) satisfies the linear system

$$\frac{d\mathbf{x}}{dt} = A\mathbf{x} \tag{5}$$

where A now is the 3×3 matrix defined by

```
» A = [-0.2    0    0.2
        0.2 -0.4    0
          0   0.4 -0.2 ]
A =
   -0.2000          0    0.2000
    0.2000    -0.4000         0
         0     0.4000   -0.2000
```

If initially Tank 1 contains 10 pounds of salt and the other two tanks contain pure water, then the initial vector $\mathbf{x}(0) = \mathbf{x}_0$ is defined by

```
» x0 = [10; 0; 0];
```

The eigenvalues and eigenvectors of A are calculated using the command

```
» [V, E] = eig(A)
V =
  -0.6667             0.4976 + 0.0486i   0.4976 - 0.0486i
  -0.3333             0.0486 - 0.4976i   0.0486 + 0.4976i
  -0.6667            -0.5462 + 0.4491i  -0.5462 - 0.4491i

E =
  -0.0000                   0                  0
        0            -0.4000 + 0.2000i         0
        0                   0          -0.4000 - 0.2000i
```

Now we see complex conjugate pairs of eigenvalues and eigenvectors, but let us nevertheless proceed without fear, hoping that "ordinary" real-valued solution functions will somehow result. First we pick off and record the eigenvalues that appear as the diagonal elements of the matrix E.

```
» L = diag(E);  L'
ans =
  -0.0000                -0.4000 - 0.2000i  -0.4000 +
0.2000i

» L1 = L(1);     L2 = L(2);     L3 = L(3);
```

The associated eigenvectors are the corresponding column vectors

```
» v1 = V(:,1);  v2 = V(:,2);  v3 = V(:,3);
```

of the matrix V. To find the constants c_1, c_2, c_3 in the solution

$$\mathbf{x}(t) = c_1\mathbf{v}_1 e^{\lambda_1 t} + c_2\mathbf{v}_2 e^{\lambda_2 t} + c_3\mathbf{v}_3 e^{\lambda_3 t} \tag{6}$$

we need only solve the system $V\mathbf{c} = \mathbf{x}_0$:

```
»  c = V\x0;   c'
ans =
   -6.0000 + 0.0000i   5.7773 + 2.5735i   5.7773 - 2.5735i
```

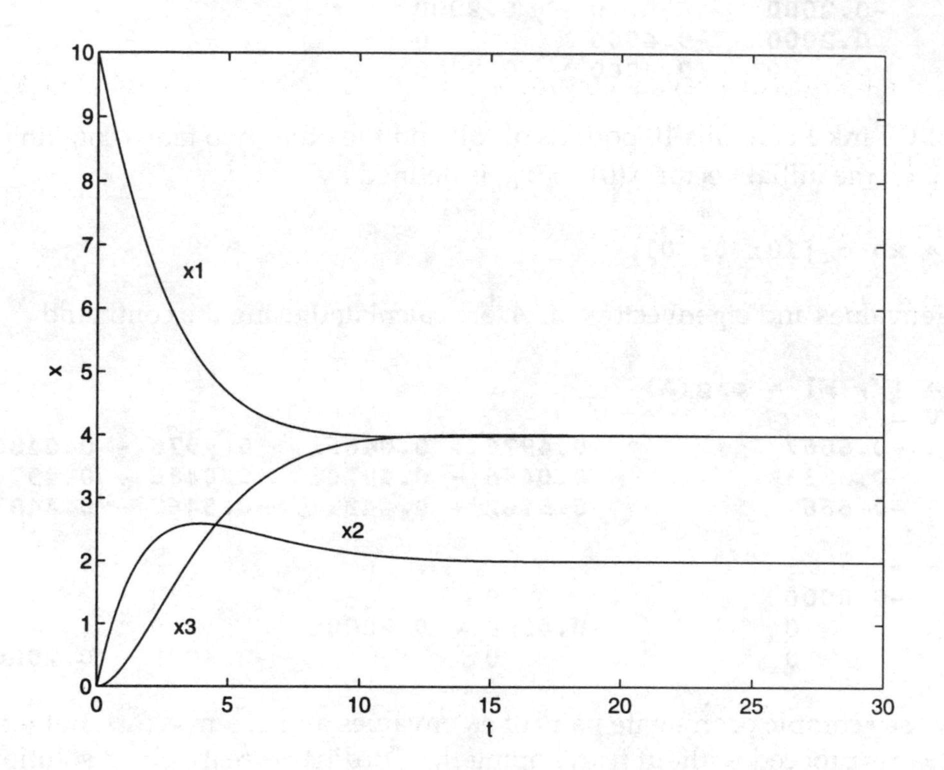

We can plot the three component functions

```
» x1 = x(1,:);    x2 = x(2,:);    x3 = x(3,:);
```

using the command

```
» plot( t,x1, t,x2, t,x3 )
```

The result is the figure above showing the graphs of the functions $x_1(t)$, $x_2(t)$, and $x_3(t)$ that give the amounts of salt in the three tanks. Is it clear to you that the three solution curves "level off" as $t \to \infty$ in a way that exhibits a long-term uniform concentration of salt throughout the system?

Project 21
Earthquake-Induced Vibrations of Multistory Buildings

Reference: Section 5.3 of Edwards & Penney
DIFFERENTIAL EQUATIONS with Computing and Modeling

In this project you are to investigate the response to transverse earthquake ground oscillations of a seven-story building like the one illustrated in Fig. 5.3.13 of the text. Suppose that each of the seven [aboveground] floors weighs 16 tons, so the mass of each is $m = 1000$ (slugs). Also assume a horizontal restoring force of $k = 5$ (tons per foot) between adjacent floors. That is, the internal forces in response to horizontal displacements of the individual floors are those shown in Fig. 5.3.14. It follows that the free transverse oscillations indicated in Fig. 5.3.13 satisfy the equation $M\mathbf{x}'' = K\mathbf{x}$ with $n = 7$ and $m_i = 1000$, $k_i = 10{,}000$ (lb/ft) for $1 \le i \le 7$. The system then reduces to the form $\mathbf{x}'' = A\mathbf{x}$ with

$$A = \begin{bmatrix} -20 & 10 & 0 & 0 & 0 & 0 & 0 \\ 10 & -20 & 10 & 0 & 0 & 0 & 0 \\ 0 & 10 & -20 & 10 & 0 & 0 & 0 \\ 0 & 0 & 10 & -20 & 10 & 0 & 0 \\ 0 & 0 & 0 & 10 & -20 & 10 & 0 \\ 0 & 0 & 0 & 0 & 10 & -20 & 10 \\ 0 & 0 & 0 & 0 & 0 & 10 & -10 \end{bmatrix} \tag{1}$$

Once the matrix A has been entered, the TI-85 command **`eigVl A`** takes less than 15 seconds to calculate the seven eigenvalues shown in the λ-row of the table below. Alternatively, you can use the *Maple* **`with(linalg)`** command **`eigenvals(A)`**, the *Mathematica* command **`Eigenvalues[A]`**, or the MATLAB command **`eig(A)`**. Then the remaining entries $\omega = \sqrt{-\lambda}$ and $P = 2\pi / \omega$ showing the natural frequencies and periods of oscillation of the seven-story building are readily calculated. Note that a typical earthquake producing ground oscillations with a period of 2 seconds is uncomfortably close to the fifth natural frequency 1.9869 seconds of the building.

i	1	2	3	4	5	6	7
λ	-38.2709	-33.3826	-26.1803	-17.9094	10.0000	-3.8197	-0.4370
ω	6.1863	5.7778	5.1167	4.2320	3.1623	1.9544	0.6611
P (sec)	1.1057	1.0875	1.2280	1.4847	1.9869	3.2149	9.5042

A horizontal earthquake oscillation $E \cos \omega t$ of the ground, with amplitude E and acceleration $a = -E\omega^2 \cos \omega t$, produces an opposite inertial force $F = ma = mE\omega^2 \cos \omega t$ on each floor of the building. The resulting nonhomogeneous linear system is

$$\mathbf{x}'' = A\mathbf{x} + (E\omega^2 \cos \omega t)\,\mathbf{b} \tag{2}$$

where $\mathbf{b} = [1 \;\; 1 \;\; 1 \;\; 1 \;\; 1 \;\; 1 \;\; 1]^T$ and A is the matrix of Eq. (1). Figure 5.3.16 in the text shows a plot of *maximal amplitude* (for the forced oscillations of any single floor) versus the *period* of the earthquake vibrations. The spikes correspond to the first six of the seven resonance frequencies. We see, for instance, that while an earthquake with period 2 seconds likely would produce destructive resonance vibrations in the building, it probably would be unharmed by an earthquake with period 2.5 seconds. Different buildings have different natural frequencies of vibration, and so a given earthquake may demolish one building, but leave untouched the one next door. This type of apparent anomaly was observed in Mexico City after the devastating earthquake of September 19, 1985.

For your personal seven-story building to investigate, let the weight (in tons) of each story equal the largest digit of your student ID number and let k (in tons/ft) equal the smallest nonzero digit. Produce numerical and graphical results like those illustrated in Figs. 5.3.15 and 5.3.16 of the text. Is your building susceptible to likely damage from an earthquake with period in the 2 to 3 second range?

You might like to begin by working manually the following warm-up problems.

1. Find the periods of the natural vibrations of a building with two aboveground floors, with each weighing 16 tons and with each restoring force being $k = 5$ tons/ft.

2. Find the periods of the natural vibrations of a building with three aboveground floors, with each weighing 16 tons and with each restoring force being $k = 5$ tons/ft.

3. Find the natural frequencies and natural modes of vibration of a building with three aboveground floors as in Problem 20 in Section 5.3 of the text, except that the upper two floors weigh 8 tons instead of 16 tons. Give the ratios of the amplitudes $A, B,$ and C of oscillations of the three floors in the form $A : B : C$ with $A = 1$.

4. Suppose that the building of Problem 3 is subject to an earthquake in which the ground undergoes horizontal sinusoidal oscillations with a period of

3 s and an amplitude of 3 in. Find the amplitudes of the resulting steady periodic oscillations of the three above-ground floors. Assume the fact that a motion $E \cos \omega t$ of the ground, with amplitude E and acceleration $a = -E\omega^2 \cos \omega t$, produces an opposite inertial force $F = ma = mE\omega^2 \cos \omega t$ on a floor of mass m.

A Three-Mass Automobile Model

In the paragraphs that follow, we illustrate appropriate *Maple, Mathematica*, and MATLAB techniques by analyzing the natural frequencies of vibration of a car that is modeled by a system of three masses and four springs. Suppose that

- mass m_1 is connected to the chassis by spring k_1;
- masses m_1 and m_2 are connected by spring k_2;
- masses m_2 and m_3 are connected by spring k_3; and
- masses m_1 and m_3 are connected by spring k_4.

The corresponding linear system $M\mathbf{x}'' = K\mathbf{x}$ has coefficient matrices

$$M = \begin{bmatrix} m_1 & 0 & 0 \\ 0 & m_2 & 0 \\ 0 & 0 & m_3 \end{bmatrix} \tag{3}$$

and

$$K = \begin{bmatrix} -(k_1 + k_2 + k_4) & k_2 & k_4 \\ k_2 & -(k_2 + k_3) & k_3 \\ k_4 & k_3 & -(k_3 + k_4) \end{bmatrix} \tag{4}$$

The displacement vector $\mathbf{x}(t) = \begin{bmatrix} x_1(t) & x_2(t) & x_3(t) \end{bmatrix}^T$ of the three-mass system then satisfies the equation

$$\mathbf{x}'' = A\mathbf{x} \tag{5}$$

with coefficient matrix $A = M^{-1}K$.

We will use the numerical values $m_1 = 40$, $m_2 = 20$, $m_3 = 40$ (in slugs) and $k_1 = 5000$, $k_2 = 1000$, $k_3 = 2000$, $k_4 = 3000$ (in lbs/ft). We want to find the three natural frequencies ω_1, ω_2, ω_3 of oscillation of this three-mass system modeling our car. If the car is driven with velocity v (ft/sec) over a washboard surface shaped like a cosine curve with a wavelength of $a = 30$ feet, then the result is a periodic force on the car with frequency $\omega = 2\pi v / a$. We would expect the car to experience resonance vibrations when this forcing frequency equals one of the car's natural frequencies.

Using *Maple*

First we define the masses

```
> m1 := 40:    m2 := 20:    m3 := 40:
```

and the spring constants

```
> k1 := 5000:   k2 := 1000:   k3 := 2000:   k4 := 3000:
```

Then the mass, stiffness, and coefficient matrices of our system are defined by

```
> with(linalg):
> M := diag(m1,m2,m3);
```

$$M := \begin{bmatrix} 40 & 0 & 0 \\ 0 & 20 & 0 \\ 0 & 0 & 40 \end{bmatrix}$$

```
> K := matrix(3,3, [-(k1+k2+k3),     k2,          k4,
                          k2,      -(k2+k3),      k3,
                          k4,         k3,     -(k3+k4)] );
```

$$K := \begin{bmatrix} -8000 & 1000 & 3000 \\ 1000 & -3000 & 2000 \\ 3000 & 2000 & -5000 \end{bmatrix}$$

```
> A := evalm( inverse(M) &* K );
```

$$A := \begin{bmatrix} -200 & 25 & 75 \\ 50 & -150 & 100 \\ 75 & 50 & -125 \end{bmatrix}$$

The eigenvalues of A are given by

```
> eigs := evalf(eigenvals(A));
```

$$eigs := -200., -27.8129452, -247.1870548$$

and we record them in increasing order of magnitude:

```
> L := matrix(1,3, [eigs[2],eigs[1],eigs[3]] );
```

$$L := [\,-27.8129452 \quad -200. \quad -247.1870548\,]$$

The corresponding natural frequencies of the system are then given by

```
> w := matrix(1,3, [sqrt(-L[1,1]), sqrt(-L[1,2]),
                    sqrt(-L[1,3])] );
```

$$w := [5.273797986 \quad 14.14213562 \quad 15.72218353]$$

When the car is driven over a washboard surface with wavelength

```
> a := 30:        # in feet
```

the resulting critical velocities are given by

```
> v := evalf(evalm( (w*a)/(2*Pi) )):
```

and — since 88 ft/sec corresponds to 60 miles per hour — by

```
> mph := evalm( 60*v/88 );
```

$$mph := [17.16854355 \quad 46.03890251 \quad 51.18265687]$$

in miles per hour. If the car is accelerated from 0 to 60 mph, it would therefore experience resonance vibrations as it passes through the speeds of 17, 46, and 51 miles per hour.

Using *Mathematica*

First we define the masses

```
m1 = 40;    m2 = 20;    m3 = 40;
```

and the spring constants

```
k1 = 5000;    k2 = 1000;    k3 = 2000;    k4 = 3000;
```

Then the mass, stiffness, and coefficient matrices of our system are defined by

```
M = DiagonalMatrix[m1,m2,m3];  M // MatrixForm

40   0    0
0    20   0
0    0    40
```

```
K = { {-(k1+k2+k3),   k2,          k4    },
      {     k2,      -(k2+k3),     k3    },
      {     k4,       k3,      -(k3+k4)} };
K // MatrixForm
```

```
-8000   1000    3000
1000    -3000   2000
3000    2000    -5000
```

```
A = Inverse[M] . K;
A // MatrixForm
```

```
-200    25      75
50      -150    100
75      50      -125
```

The eigenvalues of A are given by

```
eigs = Eigenvalues[A] // N
```

```
{-200., -247.187, -27.8129}
```

and we sort them in increasing order of magnitude:

```
L = Reverse[Sort[eigs]]
```

```
{-27.8129, -200., -247.187}
```

The corresponding natural frequencies of the system are then given by

```
w = Sqrt[-L]
```

```
{5.2738, 14.1421, 15.7222}
```

When the car is driven over a washboard surface with wavelength

```
a = 30;         (* in feet  *)
```

the resulting critical velocities are given by

```
v = w*a/(2*Pi) // N;
```

and — since 88 ft/sec corresponds to 60 miles per hour — by

```
mph = 60*v/88
```

```
{17.1685, 46.0389, 51.1827}
```

in miles per hour. If the car is accelerated from 0 to 60 mph, it would therefore experience resonance vibrations as it passes through the speeds of 17, 46, and 51 miles per hour.

Using MATLAB

First we define the masses

```
» m1 = 40;    m2 = 20;    m3 = 40;
```

and the spring constants

```
» k1 = 5000;    k2 = 1000;    k3 = 2000;    k4 = 3000;
```

Then the mass, stiffness, and coefficient matrices of our system are defined by

```
» M = diag([m1  m2  m3])
M =
    40     0     0
     0    20     0
     0     0    40

» K = [-(k1+k2+k3),     k2,          k4;
             k2,      -(k2+k3),      k3;
             k4,         k3,     -(k3+k4) ]
K =
       -8000        1000        3000
        1000       -3000        2000
        3000        2000       -5000

» A = M\K
A =
   -200    25    75
     50  -150   100
     75    50  -125
```

The eigenvalues of A are given by

```
» eigs = eig(A)'
eigs =
   -27.8129 -247.1871 -200.0000
```

and we sort them in increasing order of magnitude:

```
» L = fliplr(sort(eigs))
L =
   -27.8129 -200.0000 -247.1871
```

The corresponding natural frequencies of the system are then given by

```
» w = sqrt(-L)
w =
     5.2738    14.1421    15.7222
```

When the car is driven over a washboard surface with wavelength

```
» a = 30;        % in feet
```

the resulting critical velocities are given by

```
» v = w*a/(2*pi);
```

and — since 88 ft/sec corresponds to 60 miles per hour — by

```
» mph = 60*v/88
mph =
    17.1685    46.0389    51.1827
```

in miles per hour. If the car is accelerated from 0 to 60 mph, it would therefore experience resonance vibrations as it passes through the speeds of 17, 46, and 51 miles per hour.

Resonance Vibrations of the 7-Story Building

We describe here how MATLAB was used to generate Figure 5.3.16 in the text, showing maximal amplitude of oscillations (for any single floor) as a function the period P of the earthquake. First, the commands

```
» V = ones(1,6);
» A = 10*diag(V,1) - 20*eye(7) + 10*diag(V,-1);
» A(7,7) = -10
```

were entered to set up the coefficent matrix A of Eq. (1). If we substitute the trial solution

$$\mathbf{x} = \mathbf{v} \cos \omega t$$

(with undetermined coefficient vector $\mathbf{v}$) in Eq. (2), we get the matrix equation

$$(A + \omega^2 I)\mathbf{v} = -E\omega^2 \mathbf{b}$$

that is readily solved numerically for the amplitude vector $\mathbf{v}$ of the resulting forced vibrations of the individual floors of the building. The following MATLAB function **amp** does this and then selects the maximal amplitude of forced vibration of any single floor of the building in response to an earthquake vibration with period P.

```
function   y = amp(A,P)
E = 0.25;                % earthquake amplitude
n = size(A);
n = n(1,1);              % dimension of system
Id = eye(n);             % n by n identity matrix
b = ones(n,1);           % constant vector
k = size(P);
k = k(1,2);              % length of input vector P
y = ones(1,k);           % initialize y
for j = 1:k
    w = 2*pi/P(j);
    V = (A + w*w*Id)\(-E*w*w*b);   % solution of
    y(j) = max( abs(V) );          % linear system
    end
```

To calculate the maximal response **y** to vibrations with periods 0.01, 0.02, 0.03, ..., 4.99, 5.00 we need only define the vector **P** of periods and invoke the function **amp**.

```
» P = 0.01:0.01:5;
» y = amp(A,P);
```

A modern microcomputer solves the 500 necessary 7-by-7 linear systems in a matter of seconds. Finally, we need only

```
» plot(P, y)
```

to see our results on the screen (Fig. 5.3.16 in the text).

Project 22

Computer Analysis of an Exotic 5-Dimensional System

Reference: Section 5.4 of Edwards & Penney *DIFFERENTIAL EQUATIONS with Computing and Modeling*

The goal of this project is the solution of the linear system

$$\mathbf{x}' = A\,\mathbf{x} \qquad (1)$$

whose coefficient matrix is the exotic 5-by-5 matrix

$$A = \begin{bmatrix} -9 & 11 & -21 & 63 & -252 \\ 70 & -69 & 141 & -421 & 1684 \\ -575 & 575 & -1149 & 3451 & -13801 \\ 3891 & -3891 & 7782 & -23345 & 93365 \\ 1024 & -1024 & 2048 & -6144 & 24572 \end{bmatrix} \qquad (2)$$

that is generated by the MATLAB command **`gallery(5)`**. What is so exotic about this particular matrix? Well, enter it in your calculator or computer system of choice, and then use appropriate commands to show:

- First, the characteristic equation of A reduces to $\lambda^5 = 0$, so A has the single eigenvalue $\lambda = 0$ of multiplicity 5.
- Second, there is only a single eigenvector associated with this eigenvalue, which thus has defect 4.

To seek a chain of generalized eigenvectors, show that $A^4 \neq 0$ but $A^5 = 0$ (the 5×5 zero matrix). Hence *any* nonzero 5-vector $\mathbf{u}_1$ satisfies the equation

$$(A - \lambda I)^5 \mathbf{u}_1 = A^5 \mathbf{u}_1 = \mathbf{0}.$$

Calculate the vectors $\mathbf{u}_2 = A\mathbf{u}_1$, $\mathbf{u}_3 = A\mathbf{u}_2$, $\mathbf{u}_4 = A\mathbf{u}_3$, and $\mathbf{u}_5 = A\mathbf{u}_4$ in turn. You should find that $\mathbf{u}_5$ is nonzero, and is therefore (to within a constant multiple) the unique eigenvector $\mathbf{v}$ of the matrix A. But can this eigenvector $\mathbf{v}$ you find possibly be independent of your original choice of the starting vector $\mathbf{u}_1 \neq \mathbf{0}$? Investigate this question by repeating the process with several different choices of $\mathbf{u}_1$.

Finally, having found a length 5 chain $\{\mathbf{u}_5, \mathbf{u}_4, \mathbf{u}_3, \mathbf{u}_2, \mathbf{u}_1\}$ of generalized eigenvectors based on the (ordinary) eigenvector $\mathbf{u}_5$ associated with the single eigenvalue $\lambda = 0$ of the matrix A, write five linearly independent solutions of the 5-dimensional homogeneous linear system $\mathbf{x}' = A\mathbf{x}$.

In the paragraphs that follow we illustrate appropriate *Maple*, *Mathematica*, and MATLAB techniques to analyze the 4×4 matrix

$$A = \begin{bmatrix} 35 & -12 & 4 & 30 \\ 22 & -8 & 3 & 19 \\ -10 & 3 & 0 & -9 \\ -27 & 9 & -3 & -23 \end{bmatrix} \qquad (3)$$

of Problem 31 in Section 5.4 of the text. You can use any of the other problems there (especially Problems 23-30 and 32) to practice these techniques.

Using *Maple*

First we enter the matrix in (3):

```
> with(linalg):
> A := matrix(4,4, [ 35, -12,  4,  30,
                     22,  -8,  3,  19,
                    -10,   3,  0,  -9,
                    -27,   9, -3, -23 ] ):
```

Then we explore its characteristic polynomial, eigenvalues, and eigenvectors:

```
> charpoly(A,lambda);
```

$$\lambda^4 - 4\lambda^3 + 6\lambda^2 - 4\lambda + 1$$

(that is, $(\lambda - 1)^4$)

```
> eigenvals(A);
```

$$1, 1, 1, 1$$

```
> eigenvects(A);
```

$$[1, 4, \{[0 \;\; 1 \;\; 3 \;\; 0], [-1 \;\; 0 \;\; 1 \;\; 1]\}]$$

Thus *Maple* finds only the two independent eigenvectors

```
> w1 := matrix(4,1, [ 0,  1,  3,  0]):
  w2 := matrix(4,1, [-1,  0,  1,  1]):
```

associated with the multiplicity 4 eigenvalue $\lambda = 1$, which therefore has defect 2. To explore the situation we set up the 4×4 identity matrix and the matrix $B = A - \lambda I$:

```
> Id := diag(1,1,1,1):   B := evalm( A - Id):
```

When we calculate B^2 and B^3,

```
> B2 := evalm(B &* B);   B3 := evalm(B2 &* B);
```

We find that $B^2 \neq 0$ but $B^3 = 0$, so there should be a length 3 chain associated with $\lambda = 1$. Choosing

```
> u1 := matrix(4,1,[1,0,0,0]);
```

we calculate

```
> u2 := evalm( B &* u1);
```

$$u2 := \begin{bmatrix} 34 \\ 22 \\ -10 \\ -27 \end{bmatrix}$$

```
> u3 := evalm( B &* u2);
```

$$u3 := \begin{bmatrix} 42 \\ 7 \\ -21 \\ -42 \end{bmatrix}$$

Thus we have found the length 3 chain $\{\mathbf{u}_3, \mathbf{u}_2, \mathbf{u}_1\}$ based on the (ordinary) eigen-vector $\mathbf{u}_3$. (To reconcile this result with *Maple*'s **eigenvects** calculation, you can check that $\mathbf{u}_3 + 42\mathbf{w}_2 = 7\mathbf{w}_1$.) Consequently four linearly independent solutions of the system $\mathbf{x}' = A\,\mathbf{x}$ are given by

$$\mathbf{x}_1(t) = \mathbf{w}_1 e^t, \qquad \mathbf{x}_2(t) = \mathbf{u}_3 e^t, \qquad \mathbf{x}_3(t) = (\mathbf{u}_2 + \mathbf{u}_3 t)e^t, \qquad \text{and}$$

$$\mathbf{x}_4(t) = (\mathbf{u}_1 + \mathbf{u}_2 t + \tfrac{1}{2}\mathbf{u}_3 t^2)e^t$$

Using *Mathematica*

First we enter the matrix in (3):

```
A = { { 35, -12,  4,   30 },
      { 22,  -8,  3,   19 },
      {-10,   3,  0,   -9 },
      {-27,   9, -3,  -23 } };
```

Then we explore its characteristic polynomial, eigenvalues, and eigenvectors:

```
CharacteristicPolynomial[A, r]
```

```
            2      3    4
1 - 4 r + 6 r  - 4 r  + r
```

(that is, $(r-1)^4$)

```
Eigenvalues[A]

{1, 1, 1, 1}

Eigenvectors[A]

{{-3, -1, 0, 3}, {0, 1, 3, 0}, {0, 0, 0, 0},
 {0, 0, 0, 0}}
```

Thus *Mathematica* finds only the two independent (nonzero) eigenvectors

```
w1 = {-3,-1, 0, 3};  w2 = {0, 1, 3, 0};
```

associated with the multiplicity 4 eigenvalue $\lambda = 1$, which therefore has defect 2. To explore the situation we set up the 4×4 identity matrix and the matrix $B = A-\lambda I$:

```
Id = DiagonalMatrix[1,1,1,1];      B = A - Id;
```

When we calculate B^2 and B^3,

```
B2 = B.B;       B3 = B2.B;
```

We find that $B^2 \neq 0$ but $B^3 = 0$, so there should be a length 3 chain associated with $\lambda = 1$. Choosing

```
u1 = {{1},{0},{0},{0}}
```

we calculate

```
u2 = B.u1

{{34}, {22}, {-10}, {-27}}

u3 = B.u2

{{42}, {7}, {-21}, {-42}}
```

Thus we have found the length 3 chain $\{\mathbf{u}_3, \mathbf{u}_2, \mathbf{u}_1\}$ based on the (ordinary) eigen-vector $\mathbf{u}_3$. (To reconcile this result with *Mathematica*'s **Eigenvectors** calculation, you can check that $\mathbf{u}_3 + 14\mathbf{w}_1 = -7\mathbf{w}_2$.) Consequently four linearly independent solutions of the system $\mathbf{x}' = A\,\mathbf{x}$ are given by

$$\mathbf{x}_1(t) = \mathbf{w}_1 e^t, \qquad \mathbf{x}_2(t) = \mathbf{u}_3 e^t, \qquad \mathbf{x}_3(t) = (\mathbf{u}_2 + \mathbf{u}_3 t)e^t, \qquad \text{and}$$

$$\mathbf{x}_4(t) = (\mathbf{u}_1 + \mathbf{u}_2 t + \tfrac{1}{2}\mathbf{u}_3 t^2)e^t$$

Using MATLAB

First we enter the matrix in (3):

```
» A = [ 35   -12    4    30
        22    -8    3    19
       -10     3    0    -9
       -27     9   -3   -23 ]
```

Then we explore its characteristic polynomial, eigenvalues, and eigenvectors:

```
» charpoly(A)
ans =
x^4-4*x^3+6*x^2-4*x+1
```

(that is, $(\lambda - 1)^4$)

```
» [V,L] = eigensys(A)
V =
[ 1, 0]
[ 0, 1]
[-1, 3]
[-1, 0]

L =
[1]
[1]
[1]
[1]
```

Thus MATLAB finds only the two independent eigenvectors

```
» w1 = [1  0  -1  -1]';     w2 = [0  1  3  0]';
```

associated with the multiplicity 4 eigenvalue $\lambda = 1$, which therefore has defect 2. To explore the situation we set up the 4×4 identity matrix and the matrix $B = A - \lambda I$:

```
» Id = eye(4);          B = A - Id;
```

When we calculate B^2 and B^3,

```
» B2 = B*B;          B3 = B2*B;
```

We find that $B^2 \neq 0$ but $B^3 = 0$, so there should be a length 3 chain associated with $\lambda = 1$. Choosing

```
» u1 = [1  0  0  0]'
```

we calculate

```
» u2 = B*u1
u2 =
      34
      22
     -10
     -27

» u3 = B*u2
u3 =
      42
       7
     -21
     -42
```

Thus we have found the length 3 chain $\{\mathbf{u}_3, \mathbf{u}_2, \mathbf{u}_1\}$ based on the (ordinary) eigen-vector $\mathbf{u}_3$. (To reconcile this result with MATLAB's **eigensys** calculation, you can check that $\mathbf{u}_3 - 42\mathbf{w}_1 = 7\mathbf{w}_2$.) Consequently four linearly independent solutions of the system $\mathbf{x}' = A\,\mathbf{x}$ are given by

$$\mathbf{x}_1(t) = \mathbf{w}_1 e^t, \qquad \mathbf{x}_2(t) = \mathbf{u}_3 e^t, \qquad \mathbf{x}_3(t) = (\mathbf{u}_2 + \mathbf{u}_3 t)e^t, \qquad \text{and}$$

$$\mathbf{x}_4(t) = (\mathbf{u}_1 + \mathbf{u}_2 t + \tfrac{1}{2}\mathbf{u}_3 t^2)e^t$$

Project 23
Computer-Generated Matrix Exponentials

Reference: Section 5.5 of Edwards & Penney
DIFFERENTIAL EQUATIONS with Computing and Modeling

If A is an $n \times n$ matrix, then a computer algebra system can be used to calculate the fundamental matrix e^{At} for the homogeneous linear system $\mathbf{x}' = A\mathbf{x}$. (The success of these commands and the usefulness of the results may depend upon whether sufficiently simple expressions for the eigenvalues and eigenvectors of A can be found.) Then Theorem 2 in Section 5.5 of the text says that the solution of the initial value problem

$$\mathbf{x}' = A\mathbf{x}, \qquad \mathbf{x}(0) = \mathbf{x}_0 \tag{1}$$

is given by the (matrix) product

$$\mathbf{x}(t) \;=\; e^{At}\,\mathbf{x}_0. \tag{2}$$

In the paragraphs below we illustrate this method by using *Maple*, *Mathematica*, and MATLAB to solve the initial value problem

$$\begin{aligned} x_1' &= 13x_1 + 4x_2, & x_1(0) &= 11 \\ x_2' &= 4x_1 + 7x_2, & x_2(0) &= 23. \end{aligned} \tag{3}$$

For a three-dimensional example, use the matrix exponential to solve the initial value problem

$$\begin{aligned} x_1' &= -149x_1 - 50x_2 - 154x_3, & x_1(0) &= 17 \\ x_2' &= 537x_1 + 180x_2 + 546x_3, & x_2(0) &= 43 \\ x_3' &= -27x_1 - 9x_2 - 25x_3, & x_3(0) &= 79. \end{aligned}$$

And here's a four-dimensional problem:

$$\begin{aligned} x_1' &= 4x_1 + x_2 + x_3 + 7x_4, & x_1(0) &= 15 \\ x_2' &= x_1 + 4x_2 + 10x_3 + x_4, & x_2(0) &= 35 \\ x_3' &= x_1 + 10x_2 + 4x_3 + x_4, & x_3(0) &= 55 \\ x_4' &= 7x_1 + x_2 + x_3 + 4x_4, & x_4(0) &= 75. \end{aligned}$$

If at this point you're having too much fun with matrix exponentials to stop, make up some problems of your own. For instance, choose any homogeneous linear system appearing in this chapter and experiment with differential initial conditions. The exotic 5×5 matrix A of the Project 22 may suggest some interesting possibilities.

Using *Maple*

First we define the coefficient matrix

```
> with(linalg):  A := matrix(2,2, [13, 4, 4, 7] );
```

$$A := \begin{bmatrix} 13 & 4 \\ 4 & 7 \end{bmatrix}$$

and the initial vector

```
> x0 := matrix(2,1, [11, 23]);
```

$$x0 := \begin{bmatrix} 11 \\ 23 \end{bmatrix}$$

for the initial value problem in (3). Then a fundamental matrix for the system $\mathbf{x}' = A\mathbf{x}$ is given by

```
> fundMatrix := exponential(A,t);
```

$$fundMatrix := \begin{bmatrix} \frac{1}{5}e^{5t} + \frac{4}{5}e^{15t} & -\frac{2}{5}e^{5t} + \frac{2}{5}e^{15t} \\ -\frac{2}{5}e^{5t} + \frac{2}{5}e^{15t} & \frac{4}{5}e^{5t} + \frac{1}{5}e^{15t} \end{bmatrix}$$

Hence Theorem 2 in this section gives the particular solution

```
> solution := evalm(fundMatrix &* x0);
```

$$solution := \begin{bmatrix} -7e^{5t} + 18e^{15t} \\ 14e^{5t} + 9e^{15t} \end{bmatrix}$$

Thus the desired particular solution is given in scalar form by

$$x_1(t) := -7e^{5t} + 18e^{15t}, \qquad x_2(t) := 14e^{5t} + 9e^{15t}.$$

Using *Mathematica*

First we define the coefficient matrix

```
A = {{13, 4}, {4, 7}};    A // MatrixForm

13   4
4    7
```

and the initial vector

```
x0 = {{11}, {23}};   x0 // MatrixForm

11
23
```

for the initial value problem in (3). Then a fundamental matrix for the system $\mathbf{x}' = A\mathbf{x}$ is given by

```
fundMatrix = MatrixExp[ A t ]
```

```
  5 t        15 t        5 t        15 t
 E        4 E        -2 E        2 E
{{---- + -------, ------- + -------},
  5        5          5          5

   5 t        15 t        5 t        15 t
 -2 E        2 E        4 E        E
{------- + -------, ------ + -----}}
   5          5          5        5
```

Hence Theorem 2 in this section gives the particular solution

```
solution = fundMatrix . x0;   Simplify[solution]

       5 t          15 t          5 t         15 t
{{-7 E    + 18 E     }, {14 E    + 9 E     }}
```

Thus the desired particular solution is given in scalar form by

$$x_1(t) := -7e^{5t} + 18e^{15t}, \qquad x_2(t) := 14e^{5t} + 9e^{15t}.$$

Using MATLAB

To calculate symbolic matrix exponentials with MATLAB we must use the Symbolic Math Toolbox **maple** command to send instructions to the *Maple* kernel. First we define the coefficient matrix with

```
» maple('with(linalg)')
» maple('A := matrix(2,2, [13, 4, 4, 7])')
ans =
A := MATRIX([[13, 4], [4, 7]])
```

and the initial vector with

```
» maple('x0 := matrix(2,1, [11, 23])')
ans =
x0 := MATRIX([[11], [23]])
```

for the initial value problem in (3). Then a fundamental matrix for the system $x' = Ax$ is given by

```
» maple('fundMatrix := exponential(A,t)')
ans =
fundMatrix :=
MATRIX([[1/5*exp(5*t)+4/5*exp(15*t),
                              2/5*exp(15*t)-2/5*exp(5*t)],
        [2/5*exp(15*t)-2/5*exp(5*t),
                              4/5*exp(5*t)+1/5*exp(15*t)]])
```

Hence Theorem 2 in this section gives the particular solution

```
» solution = maple('evalm(fundMatrix &* x0)')
solution =
[-7*exp(5*t)+18*exp(15*t)]
[ 9*exp(15*t)+14*exp(5*t)]
```

Thus the desired particular solution is given in scalar form by

$$x_1(t) := -7e^{5t} + 18e^{15t}, \qquad x_2(t) := 14e^{5t} + 9e^{15t}.$$

Chapter 6

Nonlinear Systems and Phenomena

Project 24
Computer-Generated Phase Plane Portraits for Two-Dimensional Systems

Reference: Section 6.1 of Edwards & Penney
DIFFERENTIAL EQUATIONS with Computing and Modeling

Consider a first-order differential equation of the form

$$\frac{dy}{dx} = \frac{G(x,y)}{F(x,y)}, \tag{1}$$

which may be difficult or impossible to solve explicitly. Its solution curves can nevertheless be plotted as trajectories of the corresponding autonomous two-dimensional system

$$\frac{dx}{dt} = F(x,y), \qquad \frac{dy}{dt} = G(x,y). \tag{2}$$

Most ODE plotters can routinely generate phase portraits for autonomous systems. (Those appearing in Chapter 6 of the text were plotted using the **`pplane`** program accompanying John Polking's *MATLAB Manual for Ordinary Differential Equations*, Prentice Hall, 1995.)

For example, to plot solution curves for the differential equation

$$\frac{dy}{dx} = \frac{2xy - y^2}{x^2 - 2xy} \tag{3}$$

we plot trajectories of the system

$$\frac{dx}{dt} = x^2 - 2xy, \qquad \frac{dy}{dt} = 2xy - y^2. \tag{4}$$

The result is shown in Fig. 6.1.17 of the text. In the paragraphs that follow the problems below we discuss the use of *Maple*, *Mathematica*, and MATLAB to construct such phase plane portraits.

Plot similarly some solution curves for the following differential equations.

1. $$\frac{dy}{dx} = \frac{4x-5y}{2x+3y}$$

2. $$\frac{dy}{dx} = \frac{4x-5y}{2x-3y}$$

3. $$\frac{dy}{dx} = \frac{4x-3y}{2x-5y}$$

4. $$\frac{dy}{dx} = \frac{2xy}{x^2-y^2}$$

5. $$\frac{dy}{dx} = \frac{x^2+2xy}{y^2+2xy}$$

Now construct some examples of your own. Homogeneous functions like those in Problems 1 through 5 — rational functions with numerator and denominator of the same degree in x and y — work well. The differential equation

$$\frac{dy}{dx} = \frac{25x+y(1-x^2-y^2)(4-x^2-y^2)}{-25y+x(1-x^2-y^2)(4-x^2-y^2)} \tag{5}$$

of this form generalizes Example 6 in Section 6.1 of the text, but would be inconvenient to solve explicitly. Its phase portrait (Fig. 6.1.18) shows two periodic closed trajectories — the circles $r = 1$ and $r = 2$. Anyone want to try for three circles?

Using *Maple*

The **`DEtools`** package includes the **`phaseportrait`** function that can be used to construct a phase plane portrait for a 2-dimensional system of first-order differential equations. For instance, given the differential equations

```
> deq1 := diff(x(t),t) = x^2 - 2*x*y:
  deq2 := diff(y(t),t) = 2*x*y - y^2:
```

in (4), in order to plot the trajectories with initial conditions $x(0) = n,\ y(0) = n$ for $n = -4, -3, \ldots\ldots, 2, 3, 4,$ we need only enter the commands

```
> with(DEtools):
> phaseportrait([deq1,deq2], [x,y], -5..5,
            {[0,1,1],[0,2,2],[0,3,3],[0,4,4],
             [0,-1,-1],[0,-2,-2],[0,-3,-3],[0,-4,-4]},
             stepsize = 0.01 );
```

specifying the differential equations, the dependent variables, the t-range for each solution curve, a list of initial conditions of the form $[t_0, x_0, y_0]$, and the desired step size. The result is the collection of first- and third-quadrant trajectories shown in the figure below. You can investigate similarly the second- and fourth-quadrant trajectories of the system in (4).

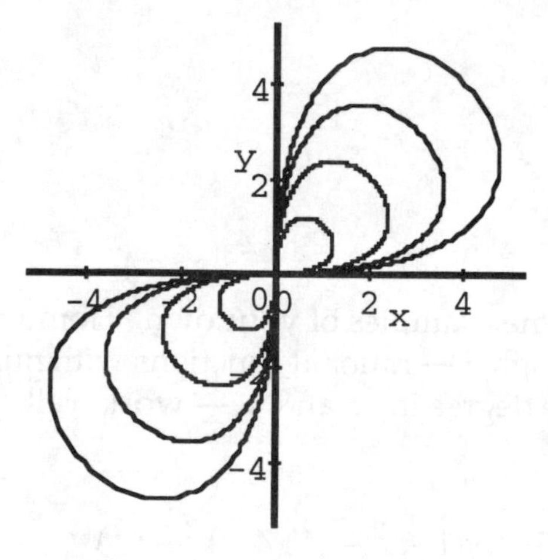

Using *Mathematica*

First we define the differential equations

```
deq1 = x'[t] == x[t]^2 - 2*x[t]*y[t];
deq2 = y'[t] == 2*x[t]*y[t] -  y[t]^2;
```

in (4). Then we can use the **NDSolve** function to solve the system numerically with given inital conditions.

```
soln = NDSolve[ {deq1, deq2, x[0] == 1, y[0] == 1},
                {x[t],y[t]},
                {t,-10,10} ]
```

```
{{x[t] -> InterpolatingFunction[{-10., 10.}, <>][t],
   y[t] -> InterpolatingFunction[{-10., 10.}, <>][t]}}
```

The corresponding solution curve is plotted with the commands

```
x = soln[[1,1,2]];  y = soln[[1,2,2]];
curve = ParametricPlot[ {x,y}, {t,-10,10},
                                AspectRatio -> 1 ];
```

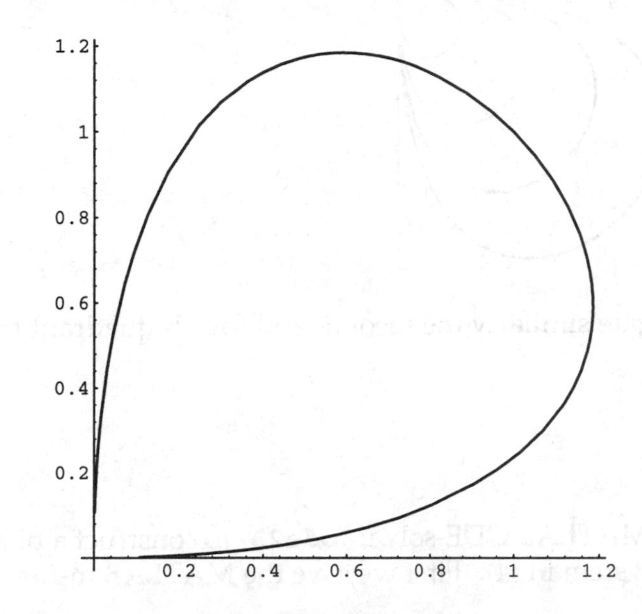

We can plot as many such solution curves as we like, and then display them simultaneously. For example, the following command creates a list (a **Table**) of solution curves corresponding to the initial conditions $x(0) = n$, $y(0) = n$ for $n =$ –4, –3, ..., 2, 3, 4.

```
curve = Table[ n, {n,-4,4} ];
Do[Clear[x,y];
   soln = NDSolve[ {deq1,deq2,x[0] == n,y[0]== n},
                   {x[t],y[t]},
                   {t,-10,10} ];
   x = soln[[1,1,2]]; y = soln[[1,2,2]];
   curve[[n]] = ParametricPlot[ {x,y}, {t,-10,10} ],
   {n,-4,4} ];
```

The command

```
Show[ curve[[-4]],curve[[-3]],curve[[-2]],curve[[-1]],
      curve[[1]],curve[[2]],curve[[3]],curve[[4]],
      AspectRatio -> 1 ];
```

then displays these solution curves in a single phase plane portrait.

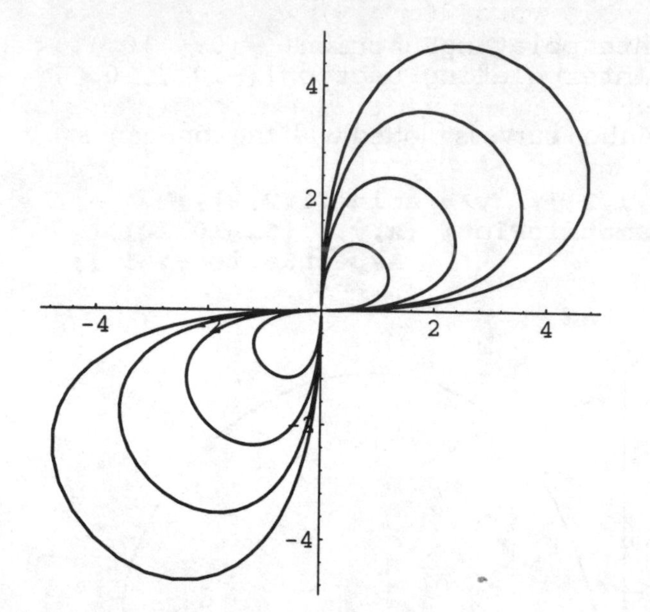

You can investigate similarly the second- and fourth-quadrant trajectories of the system in (4).

Using MATLAB

We will use the MATLAB ODE-solver **ode23** to construct a phase plane portrait for the system in (4). First we save the MATLAB m-file

```
function     yp = yproj24(t,y)
yp = y;
x = y(1);    y = y(2);
yp(1) = x.*x - 2*x.*y;
yp(2) = 2*x.*y - y.*y;
```

to define the system. Then the command

```
» [t,y] = ode23('yproj24', 0,10, [4;4] );
```

generates a 2-column matrix **y** whose two columns are the x- and y-vectors for an approximate solution of the system on the interval $0 \le t \le 10$ with initial values $x(0) = y(0) = 4$. The plot command **plot(y(:,1), y(:,2))** then produces the figure below, in which we see only half of the expected first-quadrant trajectory.

The other "half" of the desired "whole trajectory" is the solution of the system in (4) with the same initial condition, but on the interval $-10 \le t \le 0$. Because **ode23** cannot solve "backwards" with negative t-steps, we define the negative version of our original system by saving the m-file

```
function    yp = yproj24m(t,y)
yp = -yproj24(t,y);
```

We could then generate the remainder of our trajectory in similar fashion.

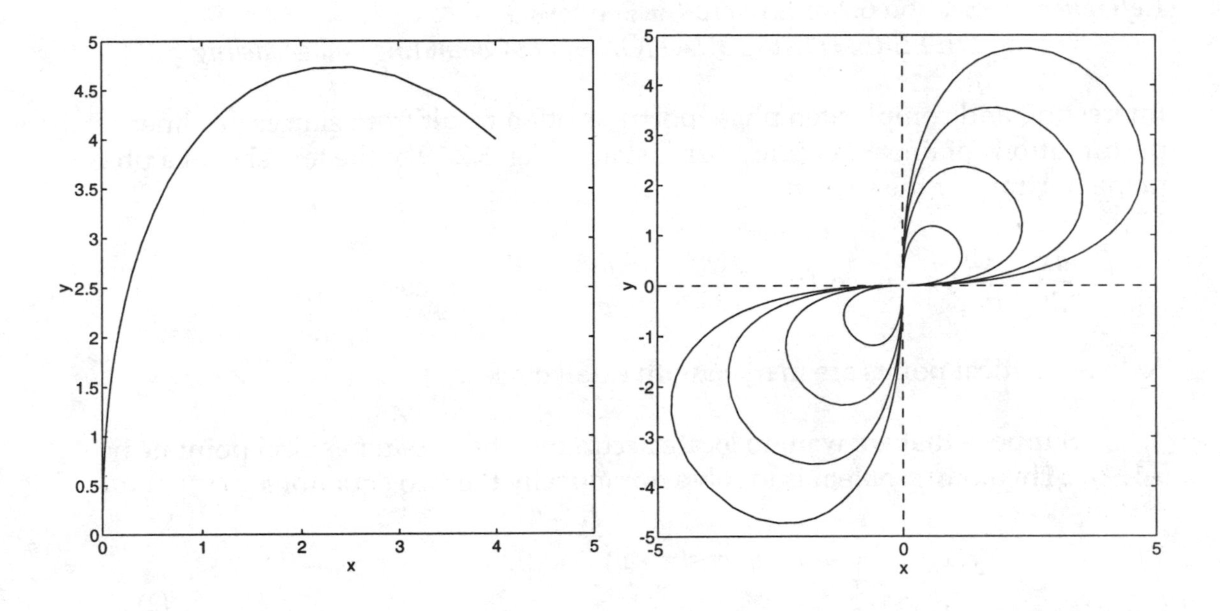

Instead, we proceed to generate directly a family of "whole" trajectories, solving the system numerically with initial conditions $x(0) = y(0) = n$ with the successive values $n = -4, -3, \ldots.., 2, 3, 4$. The figure above is generated by the simple loop

```
for n = -4 : 4
    [t,y] = ode23('yproj24', 0,10, [n;n]);
    plot( y(:,1), y(:,2) )
    hold on
    [t,y] = ode23('yproj24m', 0,10, [n;n]);
    plot( y(:,1), y(:,2) )
    end
```

in which we plot successively both "halves" of each first- and third-quadrant trajectory. You can alter the initial conditions and t-intervals specified to investigate similarly the second- and fourth-quadrant trajectories of the system in (4).

Project 25
Computer Investigation of Critical Points

Reference: Section 6.2 of Edwards & Penney
DIFFERENTIAL EQUATIONS with Computing and Modeling

Interesting and complicated phase portraits often result from simple nonlinear perturbations of linear systems. For instance, Fig. 6.2.19 in the text shows a phase plane portrait for the system

$$\frac{dx}{dt} = -x - y^2\cos(x+y), \qquad \frac{dy}{dt} = y + x^2\cos(x-y). \tag{1}$$

A dozen critical points are marked with small circles.

Suppose that we want to locate accurately the apparent spiral point near (–1, 1). Then our problem is to solve numerically the two equations

$$\begin{aligned} f(x,y) &= -x - y^2\cos(x+y) = 0, \\ g(x,y) &= y + x^2\cos(x-y) = 0 \end{aligned} \tag{2}$$

in the two unknowns x and y. In the paragraphs below we illustrate the use of *Maple*, *Mathematica*, and MATLAB for the numerical solution of nonlinear systems of equations.

Once a critical point (a, b) where $f(a,b) = g(a,b) = 0$ is located, we can classify it by looking at the Taylor expansions

$$\begin{aligned} f(x,y) &= D_xf(a,b)(x-a) + D_yf(a,b)(y-b) + \cdots \\ g(x,y) &= D_xg(a,b)(x-a) + D_yg(a,b)(y-b) + \cdots \end{aligned} \tag{3}$$

(where we have written explicitly only the terms that are linear in $x-a$ and $y-b$). The Jacobian matrix

$$J = \begin{bmatrix} D_xf(a,b) & D_yf(a,b) \\ D_xg(a,b) & D_yg(a,b) \end{bmatrix} \tag{4}$$

is the coefficient matrix of the linearization of the system (1) that results when we substitute $u = x-a,\ v = y-b$ and retain only the terms that are linear in u and v. The character of the critical point (a, b) of the original almost linear system is therefore determined (except in the ambiguous cases of Theorem 2 in Section 6.2) by the eigenvalues of the Jacobian matrix in (4).

You can proceed to investigate similarly any of the other critical points indicated in Fig. 6.2.19. For a second problem, investigate some of the critical points indicated in the Fig. 6.2.20 phase plane portrait for the system

$$\frac{dx}{dt} = -x + y\sin y, \qquad \frac{dy}{dt} = y - x^2\cos x.$$

Then investigate your own nonlinear perturbation of some simple almost linear system. You might try a system resembling one of the following:

1. $x' = 4(x + 4y + 3x^2 - y^3), \qquad y' = -3(5x - 2y + 2x^2 - y^2)$
2. $x' = x\cos y, \qquad y' = y\sin x$
3. $x' = -y + y^2\cos y, \qquad y' = -x - x^2\sin x$
4. $x' = y\cos(2x + y), \qquad y' = -x\sin(x - 3y)$
5. $x' = -y\cos(x^2 + y^2), \qquad y' = x\cos(x^2 - y^2)$

Using *Maple*

After we enter the right-hand sides in (2),

```
> f = -x - (y^2)*cos(x+y):
> g = y + (x^2)*cos(x-y):
```

we can proceed to solve numerically for a solution near $(-1, 1)$.

```
> fsolve({f=0, g=0}, {x,y},
          x=-1.5..-0.5, y=0.5..1.5);
```

$$\{y = 1.1122554536,\ x = -1.249921394\}$$

Thus our critical point (a, b) is given approximately by

```
> a := -1.24992:    b := 1.12255:
```

To classify this critical point, we proceed to set up the Jacobian matrix in (4).

```
> fx := evalf(subs(x=a,y=b,diff(f,x))):
  fy := evalf(subs(x=a,y=b,diff(f,y))):
  gx := evalf(subs(x=a,y=b,diff(g,x))):
  gy := evalf(subs(x=a,y=b,diff(g,y))):
```

```
> with(linalg):
  J:=matrix(2,2, [fx,fy,gx,gy]);
```

and to calculate its eigenvalues.

```
> eigenvals(J);
```

$$-.6233276590 + 2.567691700\ I, -.6233276590 - 2.567691700\ I$$

Thus the eigenvalues are complex conjugates with negative real parts, so the critical point (-1.24992, 1.12255) is, indeed, a stable spiral point.

Using *Mathematica*

After we enter the right-hand sides in (2),

```
f = -x - (y^2) Cos[x + y];
g = y + (x^2) Cos[x - y];
```

we can proceed to solve numerically for a solution near $(-1, 1)$.

```
FindRoot[ {f == 0, g == 0}, {x,-1}, {y,1} ]
```

```
{x -> -1.24992, y -> 1.12255}
```

Thus our critical point (a, b) is given approximately by

```
a = -1.24992;    b = 1.12255;
```

To classify this critical point, we proceed to set up the Jacobian matrix in (4).

```
J = { {D[f,x], D[f,y]},
      {D[g,x], D[g,y]}} /. {x->a, y->b};
```

and to calculate its eigenvalues.

```
Eigenvalues[J]
```

```
{-0.623328 + 2.56769 I, -0.623328 - 2.56769 I}
```

Thus the eigenvalues are complex conjugates with negative real parts, so the critical point (-1.24992, 1.12255) is, indeed, a stable spiral point.

Using MATLAB

We want to solve the equations $f(x,y) = 0$, $g(x,y) = 0$ in (2), but the student edition of MATLAB does not include a function for the solution of systems of equations. Our strategy is therefore to minimize the function

$$h(x,y) = f(x,y)^2 + g(x,y)^2 = \left(-x - y^2\cos(x+y)\right)^2 + \left(y + x^2\cos(x-y)\right)^2$$

that is defined by the m-file

```
function   z = h(v)
x = v(1);  y = v(2);
z = (-x-(y^2)*cos(x+y))^2 + (y+(x^2)*cos(x-y))^2;
```

Evidently a minimal point where $h(x,y) = 0$ will be a zero of the system in (2). Hence the command

```
» fmins('h', [-1 1])
ans =
   -1.2499     1.1225
```

shows that our critical point (a, b) near (–1, 1) is given approximately by

```
a = -1.2499;    b = 1.1225;
```

To classify this critical point, we proceed to set up the Jacobian matrix in (4). First we define the functions

```
» f = '-x - (y^2)*cos(x+y)';
» g = 'y + (x^2)*cos(x-y)';
```

and calculate their partial derivatives at (a, b):

```
» fx = diff(f,'x');      fy = diff(f,'y');
» gx = diff(g,'x');      gy = diff(g,'y');

» fxa = subs(fx,a,'x'); fxab = numeric(subs(fxa,b,'y'));
» fya = subs(fy,a,'x'); fyab = numeric(subs(fya,b,'y'));
» gxa = subs(gx,a,'x'); gxab = numeric(subs(gxa,b,'y'));
» gya = subs(gy,a,'x'); gyab = numeric(subs(gya,b,'y'));
```

Then the eigenvalues of the Jacobian matrix

```
» J = [fxab   fyab
       gxab   gyab];
```

are given by

```
» eig(J)
ans =
   -0.6234 + 2.5676i
   -0.6234 - 2.5676i
```

Because these eigenvalues are complex conjugates with negative real parts, it follows that the critical point (-1.24992, 1.12255) is, indeed, a stable spiral point.

Project 26
Periods of Predator-Prey Systems

Reference: Section 6.3 of Edwards & Penney
DIFFERENTIAL EQUATIONS with Computing and Modeling

The closed trajectories in the first figure below represent periodic solutions of the predator-prey system

$$\frac{dx}{dt} = x - xy, \qquad \frac{dy}{dt} = -y + xy, \tag{1}$$

but provide no information as to the actual periods of the oscillations these solutions describe. The period of a particular solution $(x(t), y(t))$ can be gleaned from the graphs of x and y as functions of t. The second figure below shows these graphs for the solution of (1) with initial conditions $x(0) = 1$, $y(0) = 3$.

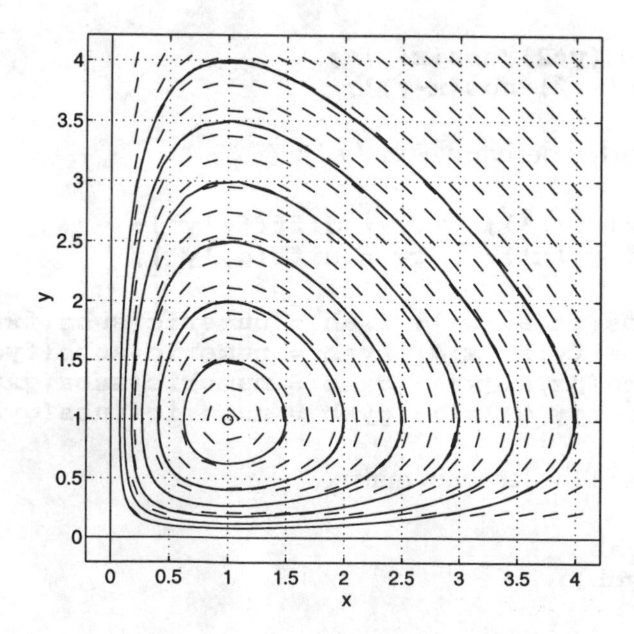

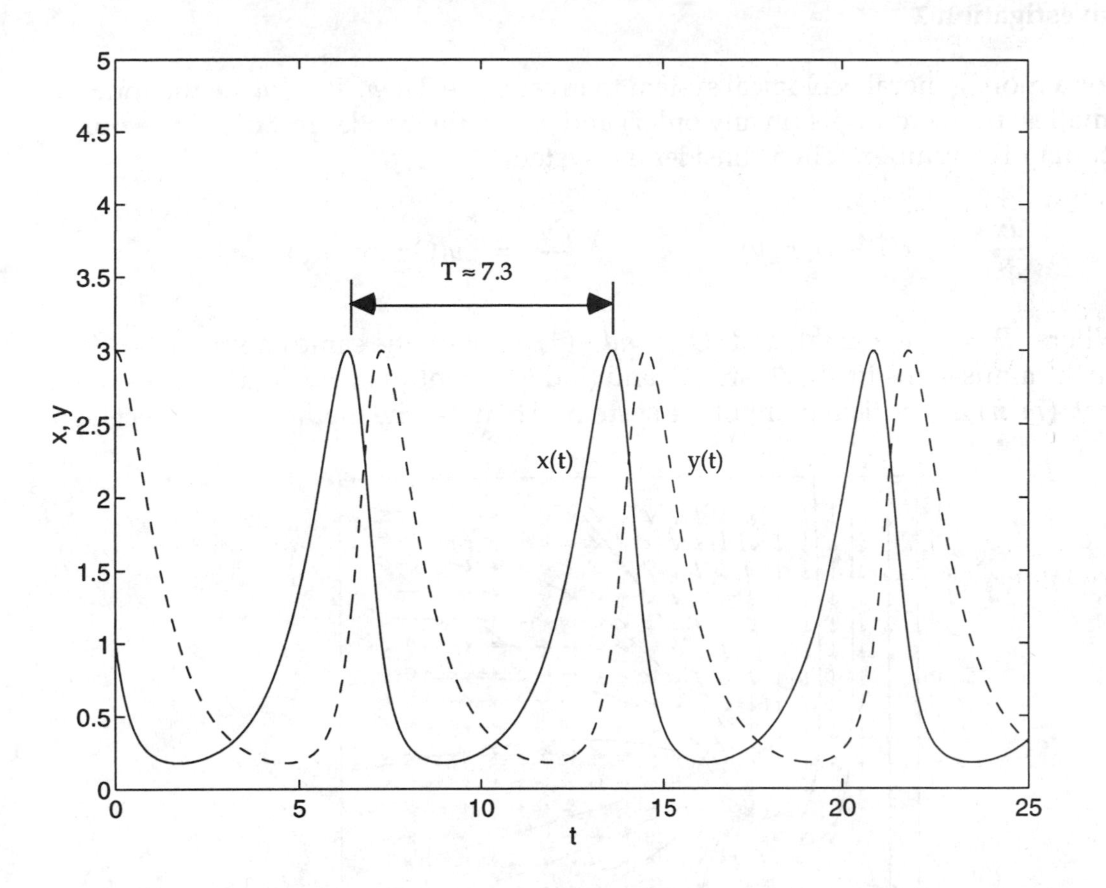

The indicated measurement on the figure indicates that the (equal) periods of oscillation of $x(t)$ and $y(t)$ are given approximately by $T \approx 7.3$.

Investigation 1

The paragraphs below illustrate *Maple*, *Mathematica*, and MATLAB commands for the construction of figures like those shown above. For your own predator-prey system

$$\frac{dx}{dt} = ax - bxy, \qquad \frac{dy}{dt} = -cy + dxy, \tag{2}$$

to investigate, let a, b, c, and d be the last four nonzero digits of your student ID number. Plot typical xy-, tx-, and ty-solution curves for this system. Measure the period of oscillation of a typical particular solution, as well as the maximal and minimal predator and prey populations.

Investigation 2

For a more general ecological system to investigate, let $a,\ b,\ c,\ d$ be the four smallest nonzero digits (in any order) and $m,\ n$ the two largest digits in your student ID number. Then consider the system

$$\frac{dx}{dt} = x(P - ax \pm by), \qquad \frac{dy}{dt} = y(Q \pm cx - dy) \tag{3}$$

Where $P = ma - (\pm nb)$ and $Q = nd - (\pm mc)$, with the same choice of plus/minus signs in dx/dt and P and (independently) in dy/dt and Q — so that (m, n) is a critical point of the system. Then use the methods of Project 24 to

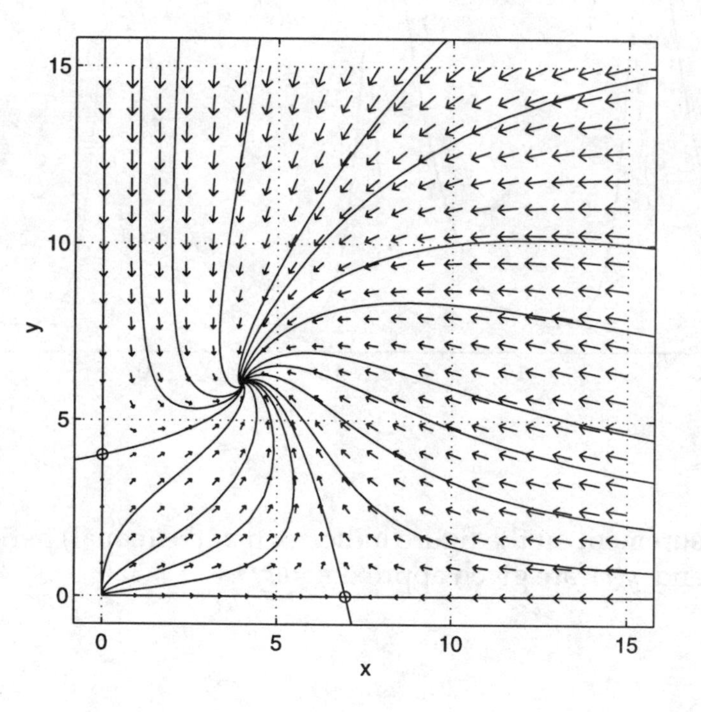

plot a phase plane portrait for this system in the first quadrant of the xy-plane. In particular, determine the long-term behavior (as $t \to \infty$) of the two populations, in terms of their initial populations $x(0) = x_0$ and $y(0) = y_0$. For instance, the figure above shows a phase plane portrait for the system

$$\frac{dx}{dt} = x(14 - 2x - y), \qquad \frac{dy}{dt} = y(8 + x - 2y).$$

We see a nodal source at $(0, 0)$, a spiral sink at $(4, 6)$, and saddle points at $(7, 0)$ and $(0, 4)$. It follows that, if x_0 and y_0 are both positive, then $x(t) \to 4$ and $y(t) \to 6$ as $t \to \infty$.

Using *Maple*

To plot a solution curve for the system in (1) we need only load the **DEtools** package and use the **DEplot2** function. For instance, the command

```
> with(DEtools):
> DEplot2([x-x*y,-y+x*y], [x,y], 0..25, {[0,1,3]},
               stepsize=0.1, arrows='NONE');
```

plots the *xy*-solution curve with initial conditions $x(0) = 1$, $y(0) = 3$ on the interval $0 \le t \le 25$ with step size $h = 0.1$. The command

```
> DEplot2([x-x*y,-y+x*y], [x,y], 0..25, {[0,1,3]},
               stepsize=0.1, arrows='NONE',
               scene = [t,x]);
```

plots the corresponding *tx*-solution curve, on which the approximate period of oscillation of the prey population can be measured.

Using *Mathematica*

To plot a solution curve for the system in (1) we need only define the differential equations

```
deq1 = x'[t] ==  x[t] - x[t]*y[t];
deq2 = y'[t] == -y[t] + x[t]*y[t];
```

and then use **NDSolve** to integrate numerically. For instance, the command

```
soln = NDSolve[ {deq1,deq2, x[0]==1, y[0]==3},
                {x[t],y[t]}, {t,0,25} ]
```

yields an approximate solution on the interval $0 \le t \le 25$ satifsfying the initial conditions $x(0) = 1$, $y(0) = 3$. Then the command

```
ParametricPlot[ Evaluate[ {x[t],y[t]} /. soln ],
                         {t,0,25} ]
```

plots the corresponding *xy*-solution curve, and the command

```
Plot[ Evaluate[ x[t] /. soln ], {t,0,25} ]
```

plots the corresponding *tx*-solution curve, on which the approximate period of oscillation of the prey population can be measured.

To plot a solution curve for the system in (1) we need only define the system by means of the m-file

```
function  xp = predprey(t,x)
xp = x;
y = x(2);     x = x(1);
xp(1) = x - x.*y;
xp(2) = -y + x.*y;
```

and then use **ode23** to integrate numerically. For instance, the command

```
» [t,x] = ode23('predprey', 0,25, [1;3]);
```

yields an approximate solution on the interval $0 \le t \le 25$ satifsfying the initial conditions $x(0) = 1$, $y(0) = 3$. Then the command

```
» plot(x(:,1), x(:,2))
```

plots the corresponding xy-solution curve, and the command

```
» plot(t, x(:,1))
```

plots the corresponding tx-solution curve, on which the approximate period of oscillation of the prey population can be measured.

Project 27

Applications of the Rayleigh and van der Pol Equations

Reference: Section 6.4 of Edwards & Penney
DIFFERENTIAL EQUATIONS with Computing and Modeling

The British mathematical physicist Lord Rayleigh (John William Strutt, 1842–1919) introduced an equation of the form

$$mx'' + kx = ax' - b(x')^3 \tag{1}$$

(with non-linear velocity damping) to model the oscillations of a clarinet reed. With $y = x'$ we get the autonomous system

$$x' = y, \qquad y' = \frac{1}{m}\left(-kx + ay - by^3\right) \tag{2}$$

whose phase plane portrait is shown in the figure below (for the case $m = k = a = b = 1$). The outward and inward spiral trajectories converge to a "limit cycle" solution that corresponds to periodic oscillations of the reed. The period T (and hence the frequency) of these oscillations can be measured on a tx-solution curve (see Fig. 6.4.17 in the text) plotted using the commands illustrated in Project 26. This period of oscillation depends only on the parameters m, k, a, and b in Eq. (1), and is independent of the initial conditions (why?).

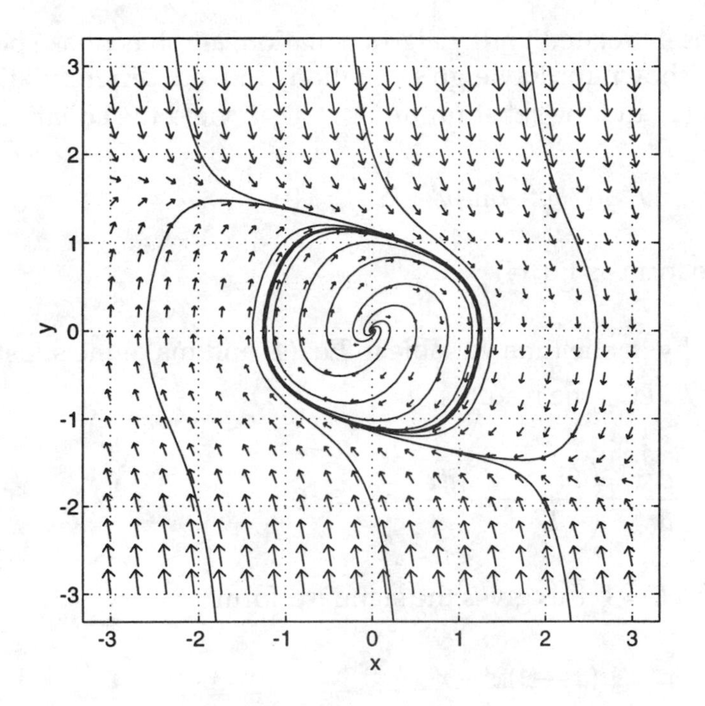

Choose your own parameters m, k, a, and b (perhaps the least four nonzero digits in your student ID number), and use an available ODE plotting utility to plot both xy-trajectories and tx-solution curves for the resulting Rayleigh equation. Change *one* of your parameters to see how the amplitude and frequency of the resulting periodic oscillations are altered.

Van der Pol's Equation

Figure 6.4.18 in the text shows a simple RLC circuit in which the usual (passive) resistance R has been replaced by an active element (such as a vacuum tube or semiconductor) across which the voltage drop V is given by a known function $f(t)$ of the current. Of course, $V = f(t) = RI$ for a standard resistor. If we substitute $f(t)$ for RI in the familiar RLC-circuit equation $LI'' + RI' + I/C = 0$ of Section 3.7, we get the second-order equation

$$LI'' + f'(I) + I/C = 0. \tag{3}$$

In a 1924 study of oscillator circuits in early commercial radios, Balthasar van der Pol (1889–1959) assumed the voltage drop to be given by a nonlinear function of the form $f(I) = bI^3 - aI$, which with Eq. (3) becomes

$$LI'' + (3bI^2 - a)I' + I/C = 0. \tag{4}$$

This equation is closely related to Rayleigh's equation, and has phase portraits resembling the one shown for Rayleigh's equation. Indeed, differentiation of the second equation in (2) and the substitution $x' = y$ yield the equation

$$my'' + (3by^2 - ay)y' + ky = 0 \tag{5}$$

which has the same form as Eq. (4).

If we denote by τ the time variable in Eq. (4) and make the substitutions $x = I\sqrt{L},\ t = \tau/\sqrt{LC}$, the result is

$$\frac{d^2x}{dt^2} + \left(\frac{3bx^2}{L} - a\right)\frac{dx}{dt} + x = 0.$$

With $\mu = a$ and $2b/aL = 1$ this gives the standard form

$$x'' + \mu(x^2 - 1)x' + x = 0 \tag{6}$$

of *van der Pol's equation.* Plot phase plane portraits and tx-solution curves with several different values of the parameter μ to see how it appears to affect the frequency of the oscillations corresponding to the limit cycle.

Project 28
Periods and Pitchforks

Reference: Section 6.5 of Edwards & Penney *DIFFERENTIAL EQUATIONS with Computing and Modeling*

We list first some programs that can be used to investigate periodic cycles for the iteration function $g(x) = rx(1-x)$, as illustrated in Figures 6.5.2–6.5.5 in the text. The following BASIC program asks the user to specify the value of the parameter r and the length k of the blocks of iterates to display. After 500 initial iterations to stabilize, 5 blocks of k iterates each are displayed for inspection.

```
100 'BASIC Program PERIODS
110 '
120 'For the iteration  x = rx(1 - x)
130 '
140  DEFDBL R,X
150  INPUT "Value of r"; R
160  INPUT "Print in blocks of k  =  "; K
170  P$  =   "#.####     "
180  X  =  .5                    'Initial seed
190 '
200  FOR I  =  1 TO 500          '500 initial
210     X  =  R*X*(1 - X)        'iterations to
220  NEXT I                      'stabilize.
230 '
240  FOR  J  =  1  TO  5
250     FOR  I  =  1  TO  K
260        X  =  R*X*(1 - X)    'Final iterations
270        PRINT USING P$; X;
280     NEXT I
290     IF K <> 8 THEN PRINT
300  NEXT J
310 '
320  END
```

Some interesting values of r (and the resulting periods) to try are

r = 2.75	Period 1
r = 3.25	Period 2
r = 3.50	Period 4
r = 3.55	Period 8
r = 3.565	Period 16
r = 3.57	Chaos
r = 3.84	Period 3
r = 3.845	Period 6
r = 3.848	Period 12

Using *Maple*

To check out the period 4 iterates with $r = 3.50$, we first carry out 500 iterations to stabilize.

```
> r := 3.5:    x := 0.5:
> for i from 1 to 500 do
       x := r*x*(1-x);
       od:
```

Then we can display 2 blocks of $k = 4$ iterates each with the commands

```
> k := 4:
> for i from 1 to 2*k do
        x := r*x*(1-x);
        od;
```

x := .8749972638
x := .3828196825
x := .8269407063
x := .5008842110

x := .8749972638
x := .3828196825
x := .8269407063
x := .5008842110

Thus we see the alleged cycles of length $k = 4$.

Using *Mathematica*

As in the text, we first define the iteration function

```
g[x_] := r x (1 - x)
```

Then the concise code

```
r = 3.5;
NestList[g, 0.5, 1012];
N[Partition[Drop[%, 1000],4] // TableForm, 4]
```

```
0.5009    0.875    0.3828    0.8269
0.5009    0.875    0.3828    0.8269
0.5009    0.875    0.3828    0.8269
```

uses the **NestList** function to calculate 1012 iterates and then display the last 12 in blocks of 4, thereby exhibiting the alleged cycles of length 4.

Using MATLAB

To check out the period 4 iterates with $r = 3.50$, we first carry out 500 iterations to stabilize.

```
»r = 3.5;    x = 0.5;
»for i = 1 : 500
     x = r*x*(1-x);
     end
```

Then we can display 3 blocks of $k = 4$ iterates each with the commands

```
» k = 4;    X = [];
» for i = 1 : 3*k
      x = r*x*(1-x);
      X = [X, x];
      end
»reshape(X, k,3)'
ans =
     0.8750     0.3828     0.8269     0.5009
     0.8750     0.3828     0.8269     0.5009
     0.8750     0.3828     0.8269     0.5009
```

Thus we see the alleged cycles of length 4.

For your further investigations, use pitchfork diagrams to look for other interesting cycles, and verify their apparent periods by appropriate iterative computations. For instance, you should find a cycle with period 10 between $r = 3.60$ and $r = 3.61$, and one with period 14 between $r = 3.59$ and $r = 3.60$. Can you find cycles with period 5 and 7? If so, look for subsequent period-doubling. A single run of of a typical pitchfork program requires several hundred thousand iterations, so it will help if you have a fast computer (or one you can leave running overnight).

Pitchfork Diagram Programs

The following BASIC program can be used to plot pitchfork diagrams as shown in Figures 6.5.6–6.5.7 in the text. As written, it runs well in Borland TurboBasic, but may have to be fine-tuned to run in other dialects of BASIC.

```
100 'Program PICHFORK
110 '
120 'Exhibits the period-doubling toward chaos
130 'generated by the Verhulst iteration
140 '
150 '        x  =  rx(1 - x)
160 '
170 'as the growth parameter  r  is increased
180 'in the range from about  3  to about  4.
190 '
200  DEFDBL H,K,R,X
210  DEFINT I,J,M,N,P,Q
220  INPUT "Rmin,Rmax"; RMIN, RMAX     'Try 2.8 and 4.0
230  INPUT "Xmin,Xmax"; XMIN, XMAX     'Try  0  and  1
240 '
250  KEY OFF  : CLS
260 'SCREEN 1   :  N  =  319           'For med resolution
270  SCREEN 2   :  N  =  639           'For hi  resolution
280  M   =  200                        'Hor rows for either
290  H   =  (RMAX - RMIN)/N
300  K   =  (XMAX - XMIN)/M
```

```
310 '
320  LINE (0,0) - (N,0)              'Draws a box
330  LINE - (N,199)                  'around
340  LINE - (0,199)                  'the screen
350  LINE - (  0,0)
360 '
370  FOR P  =  1 TO 9                'Tick marks on
380      Q  =  (P*(N+1)/10) - 1      'top and bottom
390      LINE (Q,0) - (Q,5)          'of box
400      LINE (Q,195) - (Q,199)
410  NEXT P
420 '
430  FOR J  =  0 TO  N               'Jth vertical column
440      R  =  RMIN + J*H            'of pixels on screen
450      X  =  .5
460      FOR P  =  0 TO 1000         'These iterations
470          X  =  R*X*(1-X)         'to settle down.
480      NEXT P
490      FOR Q  =  0 TO 250          'These iterations
500          X  =  R*X*(1-X)         'are recorded.
510          I  =  INT((X - XMIN)/K)
520          I  =  200 - I
530          IF (0< = I) AND (I<200) THEN PSET (J,I)
540      NEXT Q
550  NEXT J
560 '
570  WHILE INKEY$  =  ""             'Press a key when
580  WEND                            'finished looking.
590  SCREEN 0  :  CLS  :  KEY ON
600  END
```

Figures 6.5.6 and 6.5.7 in the text were plotted using the following *Mathematica* program, a slight elaboration of one found on page 102 of T. Gray and J. Glynn, **Exploring Mathematics with *Mathematica***, Addison-Wesley, 1991.

```
g[x_] : =  r x (1 - x);
Clear[r];
a  =  2.8;  b  =  4.0;   (* r-range for Fig 6.5.6 *)
c  =  0;    d  =  1;     (* x-range *)
m  =  250;                    (* no of x-points *)
n  =  500;                    (* no of r-values *)
ListPlot[
     Flatten[Table[
          Transpose[{
                    Table[r, {m+1}],
                    NestList[g, Nest[g, 0.5, 2m], m]
                    }],
               {r, a, b, (b-a)/n}
                   ],
                    1],
```

```
PlotStyle -> PointSize[0.001],
PlotRange -> {{a,b},{c,d}},
AspectRatio -> 0.75,
Frame -> True,
AxesLabel -> {"r","x"} ]
```

This *Mathematica* program runs slowly, and requires a fast machine with plenty of memory to finish within a reasonable waiting time. The following MATLAB program runs much faster on a comparable computer, and is a good deal easier to understand.

```
% pitchfork diagram script

m  =  400;                       % no of r-subintervals
n  =  250;                       % no of x-subintervals

a  =  2.8;    b  =  4.0;         % r-range for 6.5.6
dr  =  (b - a)/m;
R  =  a : dr : b;                % vector of r-values
c  =  0;      d  =  1;           % x-range
dx  =  (d - c)/n;
X  =  c : dx : d;                % vector of x-values
[rr,xx]  =  meshgrid(R,X);       % r- and x-matrices of
                           % coords of grid points in rx-rect

for i  =  0 : m                  % Cycle through r-values
    r  =  a + i*dr;
    x  =  0.5;                   % Initialize x-value
    for k  =  1:500              % First 500 iterations
        x  =  r*x*(1-x);
        end
    for k  =  1:n+1
        x  =  r*x*(1-x);
      xx(k,i+1)  =  x;           % Store x-values to plot
        end
    end

plot(rr(:,1),xx(:,1),'.w')     % Plot x-coords for each r
axis([a b c d]), hold on
for k  =  2:m+1
    plot(rr(:,k),xx(:,k),'.w')
    end
```

Project 29

Period-Doubling and Chaos in Mechanical Systems

Reference: Section 6.5 of Edwards & Penney
DIFFERENTIAL EQUATIONS with Computing and Modeling

The objective of this section is the application of the DE plotting techniques of Projects 26 and 27 to the investigation of mechanical systems that exhibit the phenomenon of period-doubling as a selected system parameter is varied.

Investigation 1

Section 6.4 of the text introduces the second-order differential equation

$$mx'' + cx' + kx + \beta x^3 = 0 \qquad (1)$$

to model the free velocity-damped vibrations of a mass m on a nonlinear spring. Recall that the term kx in Eq. (1) represents the force exerted on the mass by a *linear* spring, whereas the term βx^3 represents the nonlinearity of an actual spring. We want now to discuss the *forced vibrations* that result when an external force $F(t) = F_0 \cos \omega t$ acts on the mass. With such a force adjoined to the system in Eq. (1), we obtain the **forced Duffing equation**

$$mx'' + cx' + kx + \beta x^3 = F_0 \cos \omega t \qquad (2)$$

for the displacement $x(t)$ of the mass from its equilibrium position.

If $\beta = 0$ in (2) then we have a linear equation with stable periodic solutions. To illustrate the quite different behavior of a nonlinear system, we take $k = -1$ and $m = c = k = \beta = \omega = 1$ in Eq. (2), so the differential equation is

$$x'' + x' - x + x^3 = F_0 \cos t \qquad (3)$$

As an exercise you may verify that the two critical points $(-1, 0)$ and $(1, 0)$ are stable. We want to examine the dependence of the (presumably steady periodic) response $x(t)$ upon the amplitude F_0 of the periodic external force of period 2π.

First verify that the values $F_0 = 0.60$ and $F_0 = 0.70$ yield the two figures shown below, which indicate a simple oscillation about a critical point if $F_0 = 0.60$, and an oscillation with "doubled period" if $F_0 = 0.70$. In each case the

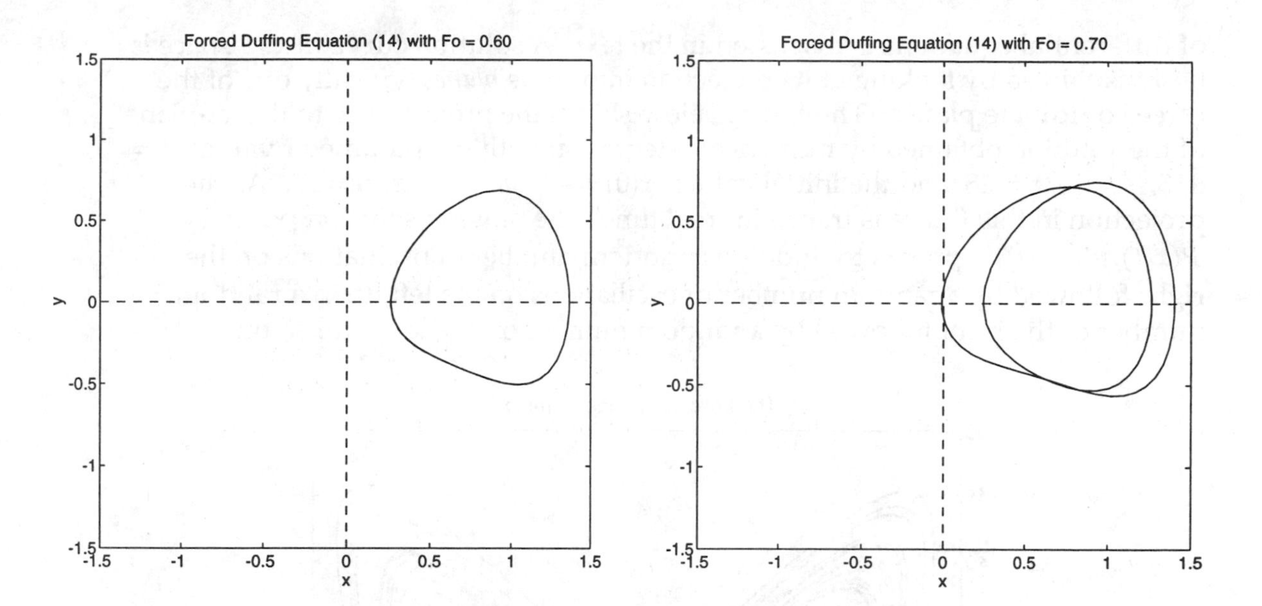

equation was solved numerically with initial conditions $x(0) = 1,\ x'(0) = 0$ and the resulting solution plotted for the range $100 \le t \le 200$ (to show the steady periodic response remaining after the initial transient response has died out). Use *tx*-plots (as in the predator-prey investigation of Project 26) to verify that the period of the oscillation with $F_0 = 0.70$ is, indeed, twice the period with $F_0 = 0.60$. Then analogous figures with $F_0 = 0.75$ and with $F_0 = 0.80$ to illustrate successive *period-doubling* and finally *chaos* as the amplitude of the external force is increased in the range from $F_0 = 0.6$ to $F_0 = 0.8$. This *period-doubling toward chaos* is a common characteristic of the behavior of a *nonlinear* mechanical system as an appropriate physical parameter (such as $m,\ c,\ k,\ \beta,\ F_0,$ or ω in Eq. (2) is increased or decreased. No such phenomenon occurs in linear systems.

Then investigate the parameter range $1.00 \le F_0 \le 1.10$ for the force constant in Eq. (3). With $F_0 = 1.00$ you should see a period 6π phase plane trajectory that encircles *both* stable critical points (as well as the unstable one). The period doubles around $F_0 = 1.07$ and chaos sets in around $F_0 = 1.10$. See whether you can spot a second period-doubling somewhere between $F_0 = 1.07$ and $F_0 = 1.10$. Produce both phase plane trajectories and *tx*-solution curves on which you can measure the periods.

Investigation 2

The genesis of the famous 3-dimensional *Lorenz system*

$$\begin{aligned} x'(t) &= -sx + sy \\ y'(t) &= -xz + rx - y \\ z'(t) &= xy - bz \end{aligned} \tag{4}$$

of differential equations is discussed in the text. A solution curve in xyz-space is best visualized by looking at its projection into some *plane*, typically one of the three coordinate planes. The figure below shows the projection into the xz-plane of the solution obtained by numerical integration with the parameter values $b = 8/3$, $s = 10, r = 28$ and the initial values $x(0) = -8$, $y(0) = 8, z(0) = 27$. As the projection in this figure is traced in "real time", the moving solution point $P(x(t), y(t), z(t))$ appears to undergo a random number of oscillations on the right followed by a random number of oscillations on the left, then a random number of the right followed by a random number on the left, and so on.

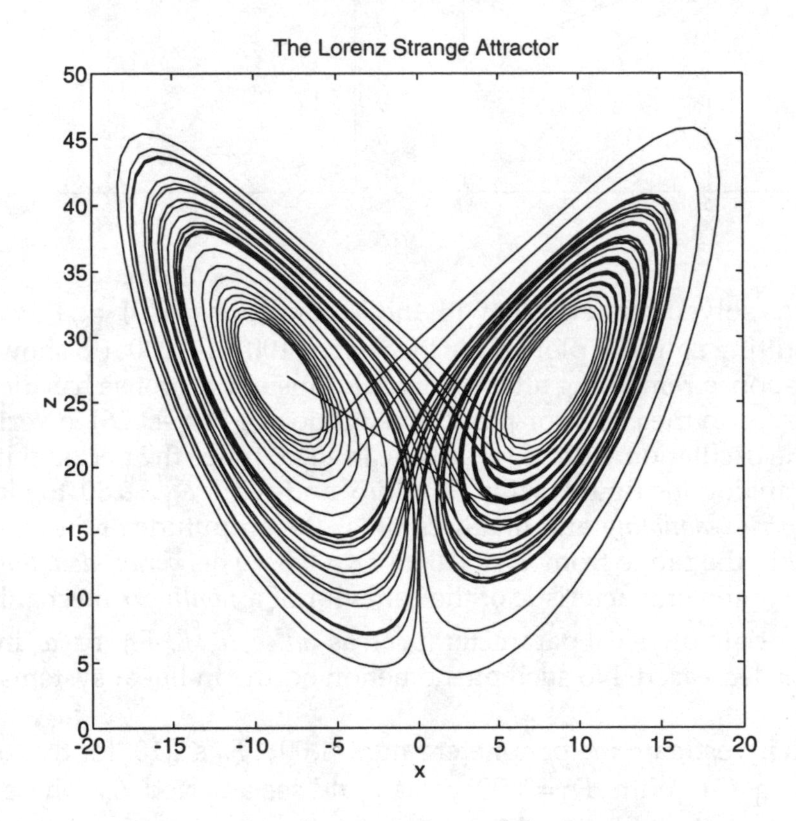

A close examination of such projections of the Lorenz trajectory shows that it is *not* simply oscillating back and forth around a pair of critical points (as the figure may initially suggest). Instead, as $t \to \infty$, the solution point $P(t)$ on the trajectory wanders back in forth in space approaching closer and closer to a certain complicated set of points whose detailed structure is not yet fully understood. This elusive set that appears somehow to "attract" the solution point is the famous *Lorenz strange attractor*.

First use an ODE plotting utility to reproduce the xz-projection of the Lorenz trajectory shown above. Use the parameter values and initial conditions listed just following Eq. (4) and numerically integrate the Lorenz system on the interval $0 \le t \le 50$. Plot also the xy- and yz-projections of this same solution. Next, experiment with different parameter values and initial conditions. For instance, see if you can find a periodic solution with $r = 70$ (and $b = 8/3$, $s = 10$ as before) and initial values $x_0 = -4$ and $z_0 = 64$. To get a trajectory that almost repeats itself, you will need to try different values of y_0 in the range $0 < y_0 < 10$ and look at xz-projections.

Investigation 3

Another much-studied nonlinear three-dimensional system is the Rossler system

$$\begin{aligned} x'(t) &= -y - z \\ y'(t) &= x + ay \\ z'(t) &= b + z(x - c) \end{aligned} \tag{5}$$

The figure below shows an xy-projection of the *Rossler band*, a chaotic attractor obtained with the values $a = 0.398$, $b = 2$, and $c = 4$ of the parameters in (5). In the xy-plane the Rossler band looks "folded," but in space it appears twisted like a Möbius strip. Investigate the period-doubling toward chaos that occurs with the Rossler system as the parameter a is increased, beginning with $a = 0.3$, $a = 0.35$, and $a = 0.375$ (take $b = 2$ and $c = 4$ in all cases).

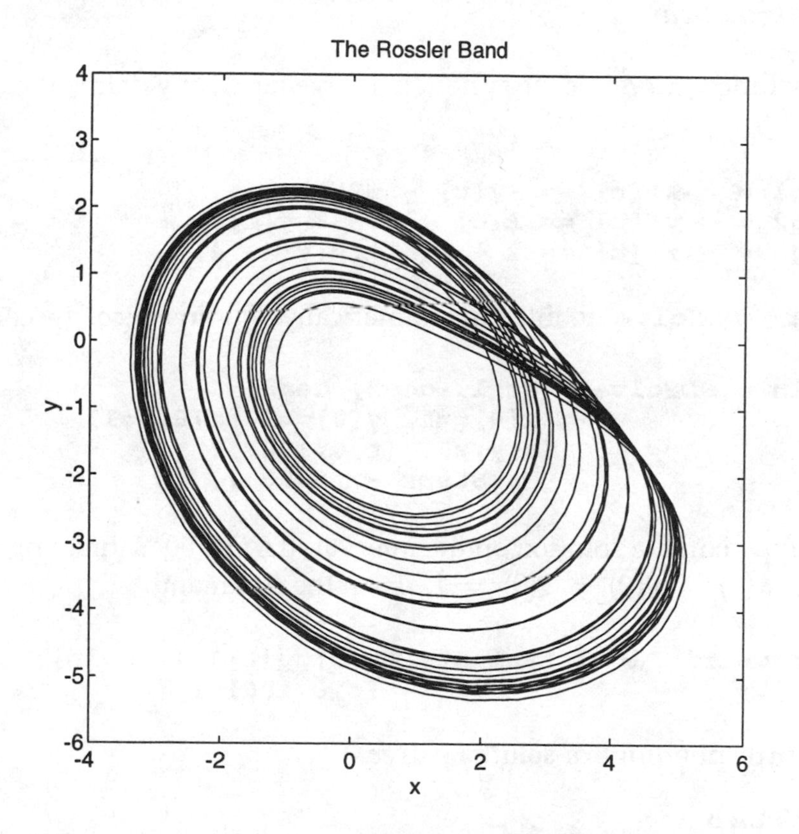

In the following paragraphs we illustrate ODE plotting techniques by showing how to use *Maple*, *Mathematica*, and MATLAB to plot the Rossler band.

Using *Maple*

To plot a solution curve for the system in (5) we need only define the system

```
> deqs := [-y-z,x+0.398*y,2+z*(x-4)]:
```

and then load the `DEtools` package and use the `DEplot` function. For instance, the command

```
> with(DEtools):
> DEplot(deqs, [t,x,y,z], 0..100, {[0,1,1,1]},
        scene = [x,y],
        stepsize = 0.1, arrows = 'NONE');
```

plots the xy-solution curve with initial conditions $x(0) = y(0) = z(0) = 1$ on the interval $0 \le t \le 100$ with step size $h = 0.1$.

Using *Mathematica*

To plot a solution curve for the system in (5) we need only define the differential equations

```
deq1 =   x'[t] == -y[t] - z[t];
deq2 =   y'[t] == x[t] + 0.398 y[t];
deq3 =   z'[t] == 2 + z[t]*(x[t] - 4);
```

and then use `NDSolve` to integrate numerically. For instance, the command

```
soln = NDSolve[ {deq1, deq2, deq3,
                 x[0]==1, y[0]==1, z[0]==1},
                {x,y,z}, {t,0,100},
                 MaxSteps -> 5000 ];
```

yields an approximate solution on the interval $0 \le t \le 100$ satifsfying the initial conditions $x(0) = y(0) = z(0) = 1$. Then the command

```
ParametricPlot[ Evaluate[{x[t],y[t]} /. soln],
                                  {t,0,100} ]
```

plots the corresponding xy-solution curve.

Using MATLAB

To plot a solution curve for the system in (5) we need only define the system by means of the m-file

```
function  xp = rossler(t,x)
xp = x;
y = x(2);    z = x(3);  x = x(1);
xp(1) = -y - z;
xp(2) = x + 0.398*y;
xp(3) = 2 + z.*(x-4);
```

and then use **ode23** to integrate numerically. For instance, the command

```
» [t,x] = [t,x] = ode23('rossler', 0,100, [1;1;1]);
```

yields an approximate solution on the interval $0 \le t \le 100$ satifsfying the initial conditions $x(0) = y(0) = z(0) = 1$. Finally, the command

```
» plot(x(:,1), x(:,2))
```

plots the corresponding xy-solution curve.

Chapter 7

Laplace Transform Methods

Project 30

Computer-Generated Laplace Transform Solutions of Linear Equations

Reference: Section 7.3 of Edwards & Penney *DIFFERENTIAL EQUATIONS with Computing and Modeling*

Most computer algebra systems can work directly with Laplace transforms. In the paragraphs that follow we illustrate the use of *Maple*, *Mathematica*, and MATLAB Laplace transform techniques in solving the mass-spring-dashpot equation

$$mx'' + cx' + kx = f(t) \tag{1}$$

with the numerical parameter values $m = 25,\ c = 10,\ k = 226$ and with selected external force functions.

Using *Maple*

Maple has the built-in functions **laplace** and **invlaplace**:

```
> laplace(t*sin(t), t, s);
```

$$2\frac{s}{\left(s^2+1\right)^2}$$

```
> invlaplace(s/(s^2 + 1)^4, s, t);
```

$$-\frac{1}{48}t^3\sin(t)-\frac{1}{16}t^2\cos(t)+\frac{1}{16}t\sin(t)$$

Free Vibrations Now let's define the mass-spring-dashpot equation in (1),

```
> deq0 := m*diff(x(t),t$2)+c*diff(x(t),t)+k*x(t)=0;
```

$$deq0 := m\left(\frac{\partial^2}{\partial t^2}x(t)\right)+c\left(\frac{\partial}{\partial t}x(t)\right)+kx(t) = 0$$

with the force function $f(t) \equiv 0$ corresponding to *free vibrations* and with the numerical parameters

```
> m:=25:    c:=10:    k:=226:
```

We can Laplace transform the whole differential equation with the command

```
> DEQ0 := laplace(deq0, t, s);
```

$$DEQ0 := 25(\text{laplace}(x(t),t,s)\,s - x(0)) - 25D(x)(0) + 10\,\text{laplace}(x(t),t,s)\,s - 10x(0) + 226\,\text{laplace}(x(t),t,s)$$

and immediately solve for the transform

```
> X(s) := solve(DEQ0, laplace(x(t), t, s));
```

$$X(s) := -\frac{-25\,s\,x(0) - 25\,D(x)(0) - 10\,x(0)}{25s^2+10s+226}$$

of a general solution of **deq0** in terms of the initial values $x(0)$ and $x'(0) = D(x)(0)$. The general solution itself is given by

```
> xc := invlaplace(X(s), s, t);
```

$$xc := \frac{1}{15}x(0)e^{-1/5\,t}\,\sin(3t) + x(0)e^{-1/5\,t}\,\cos(3t) + \frac{1}{3}D(x)(0)e^{-1/5\,t}\,\sin(3t)$$

Here we see the usual damped trigonometric oscillations of the type defined by

```
> xc := t -> exp(-t/5)*(c1*cos(3*t) + c2*sin(3*t)):
```

Forced Vibrations Now let's add a periodic driving function:

```
> deq1 :=
  m*diff(x(t),t$2)+c*diff(x(t),t)+k*x(t) = F0*cos(3*t):
```

and repeat the solution process illustrated above.

```
> DEQ1 := laplace(deq1, t,s):
> X(s) := solve(DEQ1,laplace(x(t),t,s)):
```

Next let's impose zero initial conditions

```
> X(s) := simplify(subs(x(0)=0, D(x)(0)=0, X(s)));
```

$$X(s) := \frac{F0\ s}{\left(s^2+9\right)\left(25s^2+10s+226\right)}$$

and solve for the resulting particular solution

```
> x1 := invlaplace(X(s), s, t);
```

$$x1 := \frac{30}{901} F0 \sin(3t) + \frac{1}{901} F0 \cos(3t)$$
$$- \frac{451}{13515} F0\, e^{-1/5\ t} \sin(3t) - \frac{1}{901} F0\, e^{-1/5\ t} \cos(3t)$$

Thus we see the sum of a steady periodic vibration and an exponentially damped transient solution. Let's substitute $F_0 = 901$ to plot a graph showing both the particular solution and its steady periodic part.

```
> F0 := 901:
> plot({x1, 30*sin(3*t)+cos(3*t)}, t=0..6*Pi);
```

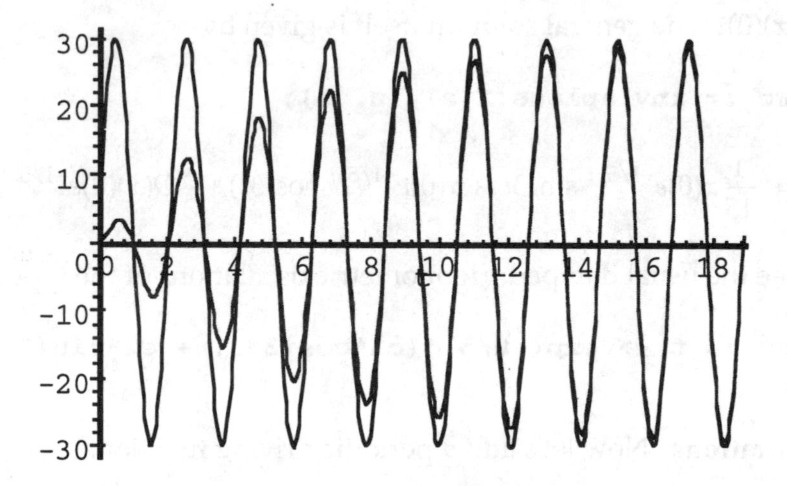

Thus we see the response gradually building up to a steady periodic oscillation.

Damped Periodic External Force Finally, let's put a damping factor in the external force:

```
> deq2 := m*diff(x(t),t$2)+c*diff(x(t),t)+k*x(t) =
                                    900*exp(-t/5)*cos(3*t):
```

and proceed as above.

```
> DEQ2 := laplace(deq2, t, s):
```

We first substitute the zero initial conditions and then inverse Laplace transform to obtain the particular solution.

```
> DEQ2 := subs(x(0)=0, D(x)(0)=0, DEQ2):
> X2(s) := simplify(solve(DEQ2, laplace(x(t),t,s)));
```

$$X(s) := 4500 \frac{5s+1}{\left(25s^2+10s+226\right)^2}$$

Here we see a repeated quadratic factor that signals the presence of a resonance phenomenon.

```
> x2 := invlaplace(X(s),s,t);
```

$$x2 := 6\,t\,e^{-1/5\,t}\sin(3t)$$

```
> plot({x2, 6*t*exp(-t/5), -6*t*exp(-t/5)}, t=0..30);
```

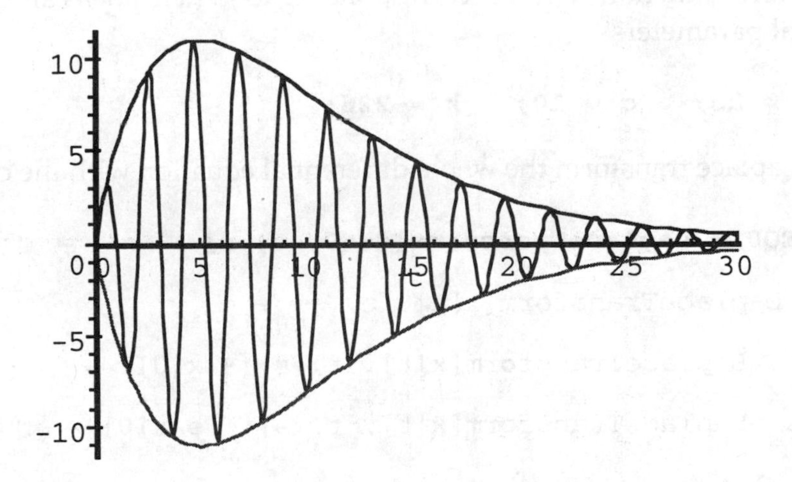

The "resonance" consists in the buildup in the amplitude of the forced oscillations before the damping prevails.

Using *Mathematica*

The functions **LaplaceTransform** and **InverseLaplaceTransform** are available in the standard **Calculus:LaplaceTransform.m** package:

```
<< Calculus`LaplaceTransform`

LaplaceTransform[ t Sin[t], t, s ]

    2 s
---------
       2 2
(1 + s )

InverseLaplaceTransform[ s/(s^2 + 1)^4, s, t ]

    2                         3
-(t  Cos[t])    t Sin[t]    t  Sin[t]
------------ + -------- - ---------
     16            16          48
```

Free Vibrations Now let's define the mass-spring-dashpot equation in (1),

```
deq0 =   m x''[t] + c x'[t] + k x[t] == 0
```

with the force function $f(t) \equiv 0$ corresponding to *free vibrations* and with the numerical parameters

```
m = 25;    c = 10;    k = 226;
```

We can Laplace transform the whole differential equation with the command

```
DEQ0 = LaplaceTransform[deq0, t, s]

226 LaplaceTransform[x[t], t, s] +

10 (s LaplaceTransform[x[t], t, s] - x[0]) +
   2
25(s  LaplaceTransform[x[t], t, s] - s x[0] - x'[0]) == 0
```

and immediately solve for the transform

```
X = Solve[DEQ0, LaplaceTransform[x[t],t,s]];
X = X[[1,1,2]]

5 (2 x[0] + 5 s x[0] + 5 x'[0])
-------------------------------
                             2
      226 + 10 s + 25 s
```

of a general solution of **deq0** in terms of the initial values $x(0)$ and $x'(0)$. The general solution itself is given by

```
xg = InverseLaplaceTransform[ X, s, t ]
```

```
    2 Sin[3 t] x[0]            Cos[3 t]    Sin[3 t]
5 (---------------- + 5 (--------  -  --------) x[0] +
           t/5                 t/5          t/5
       75 E                25 E         375 E

     Sin[3 t] x'[0]
     ---------------)
           t/5
       15 E
```

Here we see the usual damped trigonometric oscillations of the type defined by

```
xc[t_] := Exp[-t/5]*( c1 Cos[3t] + c2 Sin[3t] );
```

Forced Vibrations Now let's add a periodic driving function:

```
deq1 = m x''[t] + c x'[t] + k x[t] == F0 Cos[3t];
```

and repeat the solution process illustrated above.

```
DEQ1 = LaplaceTransform[deq1, t, s];
X = Solve[DEQ1, LaplaceTransform[x[t],t,s]];
```

Next let's impose zero initial conditions

```
X = X[[1,1,2]] /. {x[0] -> 0, x'[0] -> 0}
```

```
              F0 s
-------------------------------
       2                     2
(9 + s ) (226 + 10 s + 25 s )
```

and solve for the resulting particular solution

```
x1 = InverseLaplaceTransform[X, s, t];
x1 = Simplify[x1]
```

```
                              t/5
(F0 (-15 Cos[3 t] + 15 E    Cos[3 t] - 451 Sin[3 t] +

             t/5                           t/5
        450 E    Sin[3 t])) / (13515 E    )
```

Thus we see the sum of a steady periodic vibration and an exponentially damped transient solution. Let's substitute $F_0 = 901$ to plot the particular solution and its steady periodic part.

```
F0 = 901;
Plot[{x1, Cos[3t] + 30 Sin[3t]}, {t, 0, 6 Pi}];
```

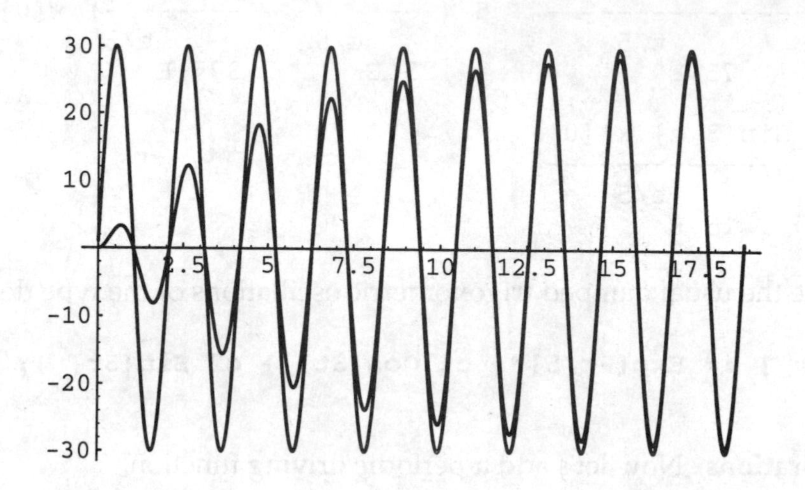

Thus we see the response gradually building up to a steady periodic oscillation.

Damped Periodic External Force Finally, let's put a damping factor in the external force:

```
deq2 =
m x''[t] + c x'[t] + k x[t] == 900 Exp[-t/5] Cos[3t];
```

and proceed as above.

```
DEQ2 = LaplaceTransform[deq2, t, s];
```

We first substitute zero initial conditions and then inverse Laplace transform to obtain the particular solution.

```
DEQ2 = DEQ2 /. {x[0] -> 0, x'[0] -> 0};
X = Solve[DEQ2, LaplaceTransform[x[t],t,s]];
X = X[[1,1,2]]

    4500 (1 + 5 s)
----------------------
                    2 2
(226 + 10 s + 25 s )
```

Here we see a repeated quadratic factor that signals the presence of a resonance phenomenon.

```
x2 = InverseLaplaceTransform[X, s, t]

6 t Sin[3 t]
------------
      t/5
     E

Plot[{x2, 6 t Exp[-t/5], -6 t Exp[-t/5]}, {t, 0,30}];
```

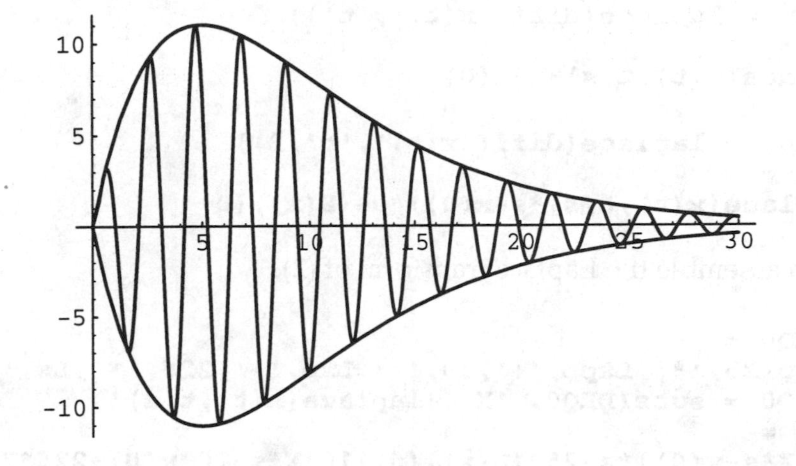

The "resonance" consists in the buildup in the amplitude of the forced oscillations before the damping prevails.

Using MATLAB

The Symbolic Math Toolbox (contained in the Student Edition) includes the functions **laplace** and **invlaplace**:

```
» laplace('t*sin(t)');  pretty

                          s
                     2 ---------
                         2     2
                       (s  + 1)

» invlaplace('s/(s^2 + 1)^4'); pretty

                3               2
       - 1/48 t  sin(t) - 1/16 t  cos(t) + 1/16 t sin(t)
```

Free Vibrations With $m = 25$, $c = 10$, $k = 226$ and with the force function $f(t) \equiv 0$ corresponding to free vibrations, the mass-spring-dashpot equation in (1) is

$$25x'' + 10x' + 226x = 0. \tag{2}$$

Let us first transform the individual terms in this equation.

```
» Lx = laplace('x(t)')
Lx =
laplace(x(t),t,s)

» Lxp = laplace(diff('x(t)','t'))
Lxp =
laplace(x(t),t,s)*s-x(0)

» Lxpp = laplace(diff('x(t)','t',2))
Lxpp =
(laplace(x(t),t,s)*s-x(0))*s-(D(x))(0)
```

Then we can assemble the Laplace transform of (2),

```
» DEQ0 =
symop(25,'*',Lxpp,'+',10,'*',Lxp,'+',226,'*',Lx,'=',0);
» DEQ0 = subs(DEQ0, 'X','laplace(x(t),t,s)')
DEQ0 =
25*(X*s-x(0))*s-25*(D(x))(0)+10*X*s-10*x(0)+226*X = 0
```

and immediately solve for the transform

```
» X = solve(DEQ0,'X');  pretty

               - 25 s x(0) - 25 D(x)(0) - 10 x(0)
        {X = - ----------------------------------}
                               2
                           25 s  + 10 s + 226
```

of a general solution of (2) in terms of the initial values $x(0)$ and $x'(0) = D(x)(0)$. The general solution itself is given by

```
» xc = invlaplace(X);    pretty

1/15 x(0) exp(- 1/5 t) sin(3 t) + x(0) exp(- 1/5 t) cos(3 t)

          + 1/3 D(x)(0) exp(- 1/5 t) sin(3 t)
```

Here we see the usual damped trigonometric oscillations of the type defined by

```
» xc = 'exp(-t/5)*(c1*cos(3*t) + c2*sin(3*t))';
```

Forced Vibrations Now let's add a periodic driving function to our mass-spring-dashpot system,

$$25\,x'' + 10\,x' + 226\,x \;=\; F_0 \cos(3t). \qquad (3)$$

The Laplace transform of the force function is

```
» F = laplace('F0*cos(3*t)')
F =
F0*s/(s^2+9)
```

so let us assemble the Laplace transform of the equation in (3) and repeat the solution process illustrated above.

```
» DEQ1 =
symop(25,'*',Lxpp,'+',10,'*',Lxp,'+',226,'*',Lx,'=',F);
» DEQ1 = subs(DEQ1, 'X','laplace(x(t),t,s)');
» X = solve(DEQ1,'X');
```

Next let's impose zero initial conditions

```
» X = subs(X, 0, 'x(0)');
» X = subs(X, 0, '(D(x))(0)');  pretty

                             F0 s
            ----------------------------
                  2             2
            (s  + 9) (25 s  + 10 s + 226)
```

and solve for the resulting particular solution

```
» x1 = invlaplace(X)
x1 =
30/901*F0*sin(3*t)+1/901*F0*cos(3*t)-
51/13515*F0*exp(-1/5*t)*sin(3*t)-
1/901*F0*exp(-1/5*t)*cos(3*t)
```

In order to see what a graph of this particular solution looks like, let's substitute the numerical value $F_0 = 901$.

```
» x1 = subs(x1,901,'F0');  pretty

                                  451
30 sin(3 t) + cos(3 t) - --- exp(- 1/5 t) sin(3 t)
                                  15

                    - exp(- 1/5 t) cos(3 t)
```

Here we see the sum of a steady periodic vibration and an exponentially damped transient solution. A plot of the particular solution, together with its steady periodic part, then shows the response gradually building up to a steady periodic oscillation.

```
» t = 0 : pi/50 : 6*pi;
» plot(t,30*sin(3*t)+cos(3*t), t,eval(vectorize(x1)))
» axis([0 6*pi -35 35]),      hold
» plot([0 6*pi],[0 0],'--')
```

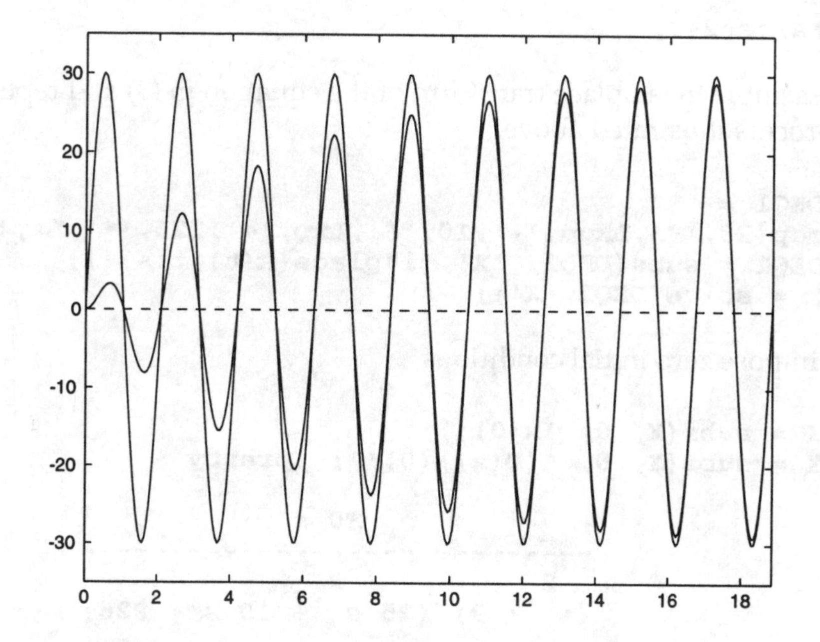

Damped Periodic External Force Finally, let's put a damping factor in the external force,

$$25\,x'' + 10\,x' + 226\,x \;=\; 900\,e^{-t/5}\,\cos(3t)$$

We calculate the transform of the new force function,

```
»F = laplace('900*exp(-t/5)*cos(3*t)')
F =
900*(s+1/5)/((s+1/5)^2+9)
```

and proceed as above.

```
» DEQ2 =
symop(25,'*',Lxpp,'+',10,'*',Lxp,'+',226,'*',Lx,'=',F);
» DEQ2 = subs(DEQ2, 'X','laplace(x(t),t,s)');
» X = solve(DEQ2,'X');
```

When we substitute the zero initial conditions,

```
» X = subs(X,0,'x(0)');
» X = subs(X, 0, '(D(x))(0)');  simple
ans =
4500*(5*s+1)/(25*s^2+10*s+226)^2
```

we see a repeated quadratic factor that signals the presence of a resonance phenomenon.

```
» x2 = invlaplace(X);  pretty

                        6 t exp(- 1/5 t) sin(3 t)

»t = 0 : 0.05 : 30;
»plot(t,eval(vectorize(x2))), hold on
»plot(t,6*t.*exp(-t/5), t,-6*t.*exp(-t/5))
»plot([0 30],[0 0],'--w')
```

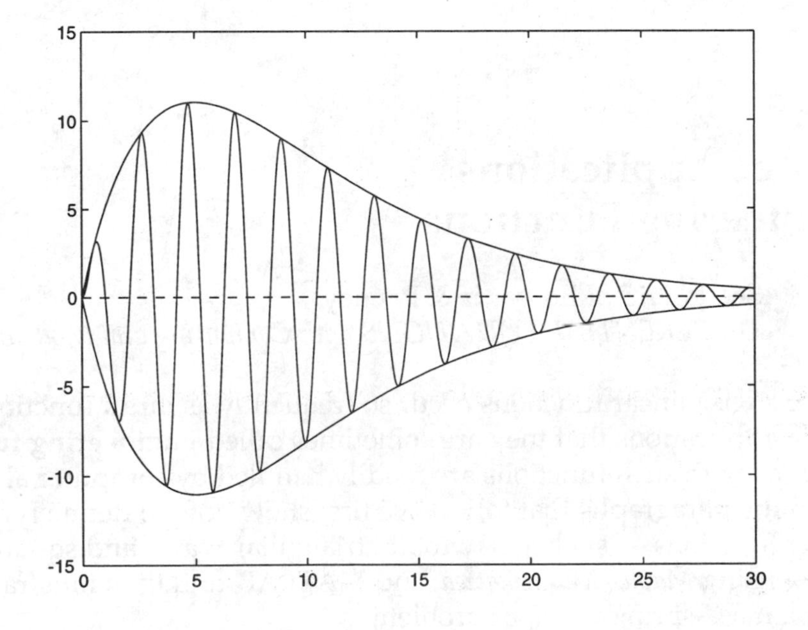

The "resonance" consists in the buildup in the amplitude of the forced oscillations before the damping prevails.

Further Investigations

You can apply similar the methods illustrated above to solve Problems 27 through 38 in Section 7.3 of the text. Better, solve some problems of your own construction. Evidently damped oscillatory force functions illustrate well the advantages of Laplace transforms. For instance,. carry out the preceding program of computations beginning with the external force function

$$f(t) \;=\; 2e^{-t/5}\,(6t\cos 3t + \sin 3t)$$

or

$$f(t) \;=\; 16200\,t^3\,e^{-t/5}\cos 3t\,,$$

and graph both the force function $f(t)$ and the response function $x(t)$. In the latter case the Laplace transform of the solution (with zero initial conditions) will exhibit a "quintic" quadratic factor that would be extremely difficult to handle manually. When you graph the resulting particular solution, you will see a beautifully slow buildup to oscillations of maximal amplitude before they are damped out. *A final question:* What force function $f(t)$ leads (with the same mass-spring-dashpot parameters and zero initial conditions as above) to the solution function $x(t) \;=\; t^3 e^{-t/5}\cos(3t)$?

Project 31
Computer Applications of Engineering Functions

Reference: Section 7.5 of Edwards & Penney
DIFFERENTIAL EQUATIONS with Computing and Modeling

Periodic piecewise linear functions occur so frequently as input functions in engineering applications that they are sometimes called **engineering functions.** Computations with such functions are readily handled by computer algebra systems. In the paragraphs that follow we first show how to define typical engineering functions — such as sawtooth, triangular-wave, and square-wave functions — using *Maple*, *Mathematica*, and MATLAB, and then illustrate the solution of a mass-spring-dashpot problem

$$m\,x'' + c\,x' + k\,x \;=\; F(t), \qquad x(0) \;=\; x'(0) \;=\; 0 \tag{1}$$

when the external force function $F(t)$ is an engineering function For your own investigation you can solve similariIy the initial value problem in (1) with various mass-spring-dashpot parameters — for instance, selected digits of you student ID number — and with input engineering functions having various amplitudes and periods.

Using *Maple*

Typical engineering functions are defined by

```
> SawTooth := t -> 2*t - 2*floor(t) - 1:

> TriangularWave :=
              t -> abs( 2*SawTooth( (2*t - 1)/4 ) ) - 1:

> SquareWave := t -> signum( TriangularWave(t) ):
```

A plot of each of these functions verifies that it has period 2 and that its name is aptly chosen. For instance, the result of the command

```
> plot( SquareWave(t), t = 0..6 );
```

shows three square-wave cycles of length 2.

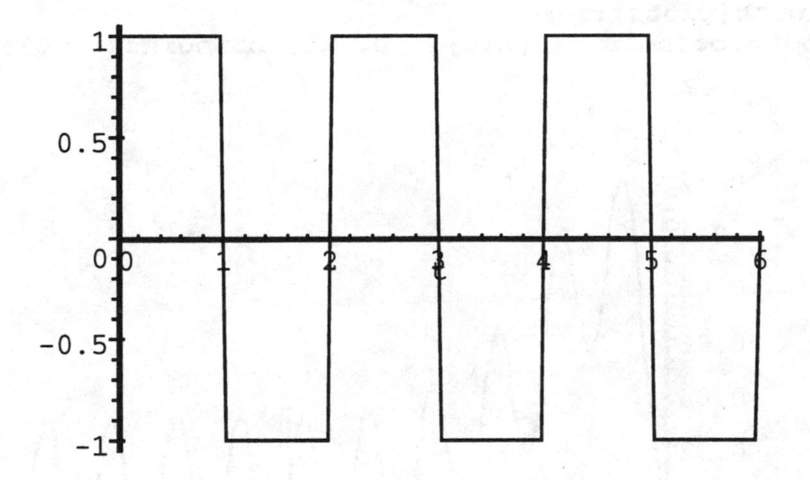

If $f(t)$ is one of the three period 2 engineering functions defined above, then the function $f(2t/p)$ will have period p. To illustrate this, try

```
> plot( TriangularWave(2*t/p), t = 0..3*p );
```

with various values of p. The resulting graph should show three triangular-wave cycles of length p.

Now let's consider the mass-spring-dashpot equation in (1) with selected parameter values and an input forcing function having period p and amplitude F_0, for instance,

```
> m := 4:      c := 8:      k := 5:
  p := 1:      F0 := 4:
  input := t -> F0*SquareWave(2*t/p);
```

You can plot this `input` function to verify that it has period 1:

```
> plot( input(t), t = 0..5 );
```

Then our differential equation is

```
> diffEq :=
     m*diff(x(t),t$2) + c*diff(x(t),t) + k*x(t) = input(t):
```

Next, let's suppose that the mass is initially at rest in its equilibrium position and solve numerically the resulting initial value problem.

```
> response :=
  dsolve( {diffEq, x(0)=0, D(x)(0)=0}, x(t), type=numeric);

> with(plots):
 odeplot(soln, [t,x(t)], 0..10, numpoints=300);
```

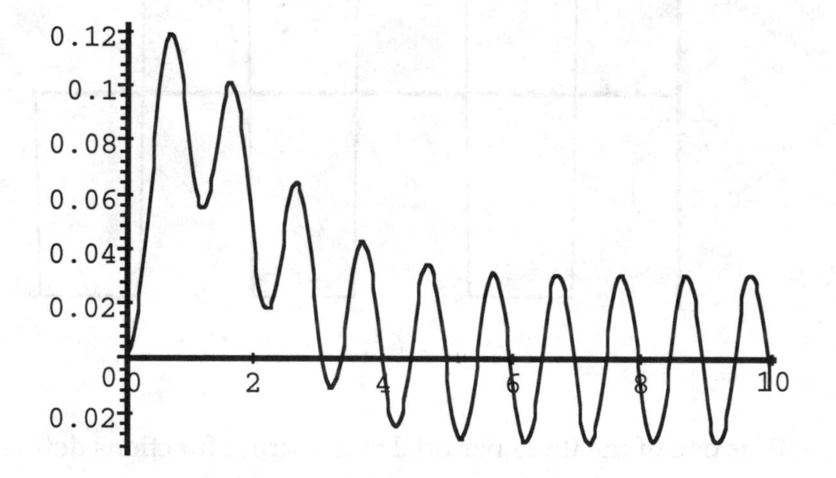

When we plot this solution, we see that after an initial transient dies out, the response function $x(t)$ settles down (as expected?) to a periodic oscillation with the same period as the input.

Using *Mathematica*

Typical engineering functions are defined by

```
SawTooth[t_] := 2 t - 2 Floor[t] - 1

TriangularWave[t_] := Abs[2 SawTooth[(2 t - 1)/4]] - 1

SquareWave[t_] := Sign[ TriangularWave[t] ]
```

A plot of each of these functions verifies that it has period 2 and that its name is aptly chosen. For instance, the result of the command

```
Plot[ SquareWave[t], {t, 0, 6} ];
```

shows three square-wave cycles of length 2.

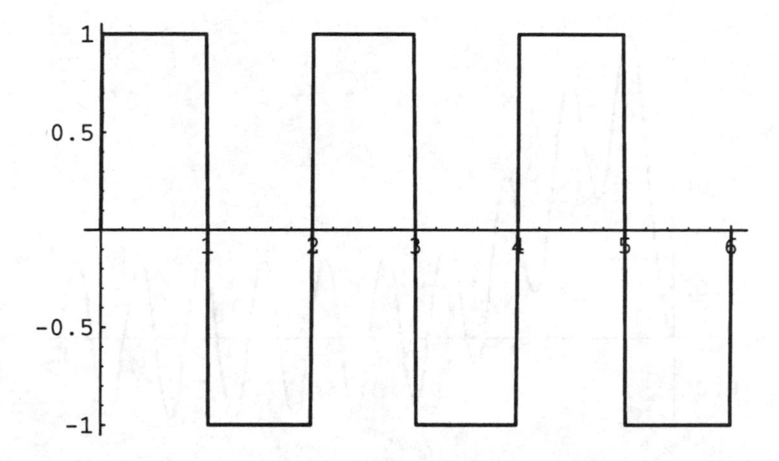

If $f(t)$ is one of the three period 2 engineering functions defined above, then the function $f(2t/p)$ will have period p. To illustrate this, try

```
Plot[ TriangularWave[ 2 t/p ], {t, 0, 3p} ];
```

with various values of p. The resulting graph should show three triangular-wave cycles of length p.

Now let's consider the mass-spring-dashpot equation in (1) with selected parameter values and an input forcing function having period p and amplitude F_0, for instance,

```
m = 4;     c = 8;     k = 5;
p = 1;     F0 = 4;
input = F0 SquareWave[2 t/p];
```

You can plot this `input` function to verify that it has period 1:

```
Plot[ input, {t, 0, 10} ];
```

Then our differential equation is

```
diffEq = m x''[t] + c x'[t] + k x[t] == input
```

Next, let's suppose that the mass is initially at rest in its equilibrium position and solve numerically the resulting initial value problem.

```
response = NDSolve[ {diffEq, x[0] == 0, x'[0] == 0},
                   x, {t, 0, 10}, MaxSteps -> 1000 ]

Plot[ x[t] /. response, {t, 0, 10} ];
```

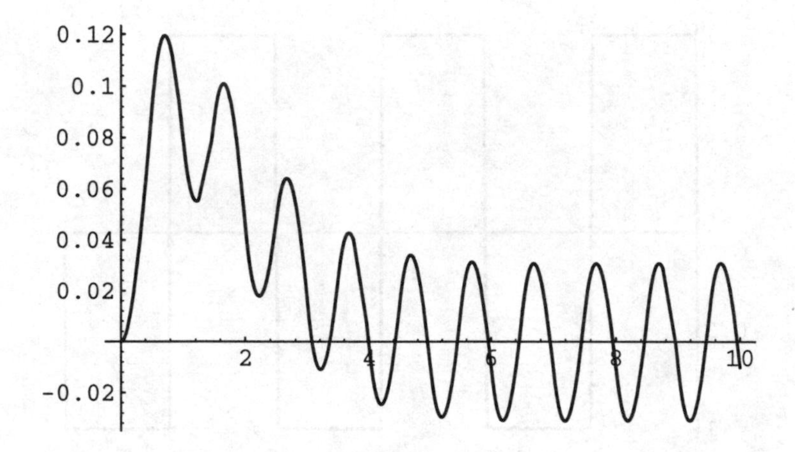

When we plot this solution, we see that after an initial transient dies out, the response function $x(t)$ settles down (as expected?) to a periodic oscillation with the same period as the input.

Using MATLAB

Typical engineering functions are defined by

```
function   x = sawtooth(t)
x = 2*t - 2*floor(t) - 1;

function    x = triangle(t)
x = abs( 2*sawtooth( (2*t - 1)/4 ) ) - 1;

function   x = square(t)
x = sign( triangle(t) );
```

A plot of each of these functions verifies that it has period 2 and that its name is aptly chosen. For instance, the result of the commands

```
» t = 0 : 0.02 : 6;        x = square(t);
» plot(t,x)
» axis([0 6 -1.2 1.2]),   hold on
» plot([0 6], [0 0])
```

shows three square-wave cycles of length 2.

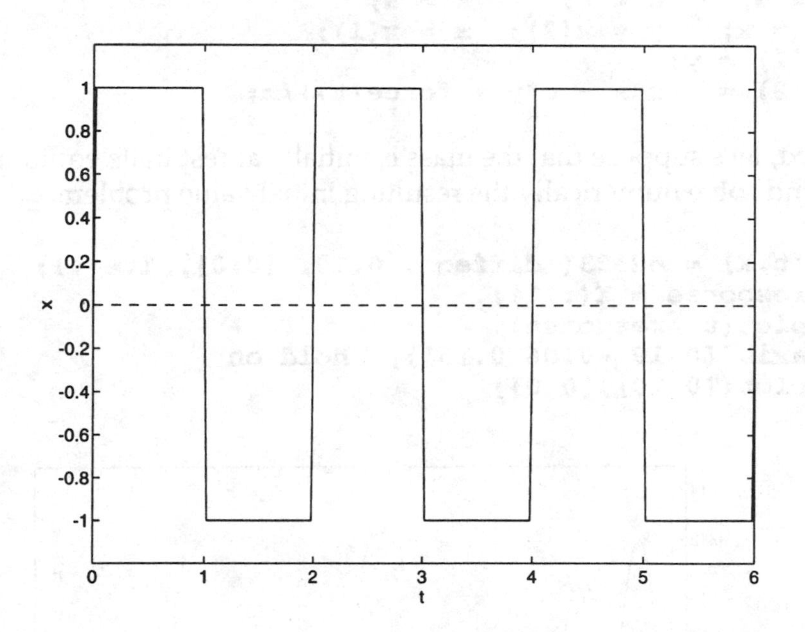

If $f(t)$ is one of the three period 2 engineering functions defined above, then the function $f(2t/p)$ will have period p. To illustrate this, try

```
» p = 4;   t = 0 : p/100 : 3*p;
» plot( t, triangle(2*t/p) )
```

with various values of p. The resulting graph should show three triangular-wave cycles of length p.

Now let's consider the mass-spring-dashpot equation in (1) with selected parameter values and an input forcing function having period p and amplitude F_0, for instance,

```
function   F = force(t)
p = 1;          % period
F0 = 4;         % amplitude
F = F0*square(2*t/p);
```

You can plot this **force** function to verify that it has period 1:

```
» t = 0 : 0.05 : 10;
» plot( t, force(t) )
```

Then our differential equation is defined (as a system of two first-order equations) by the function

```
function    xp = diffeq(t,x)
m = 4;      c = 8;        k = 5;
xp = x;     y = x(2);     x = x(1);
xp(1) = y;
xp(2) = (-k*x - c*y + force(t))/m;
```

Next, let's suppose that the mass is initially at rest in its equilibrium position and solve numerically the resulting initial value problem.

```
» [t,X] = ode23('diffeq', 0,10, [0;0], 1.e-4);
» response = X(:,1);
» plot(t, response)
» axis([0 10 -0.06 0.15]),   hold on
» plot([0 10],[0 0])
```

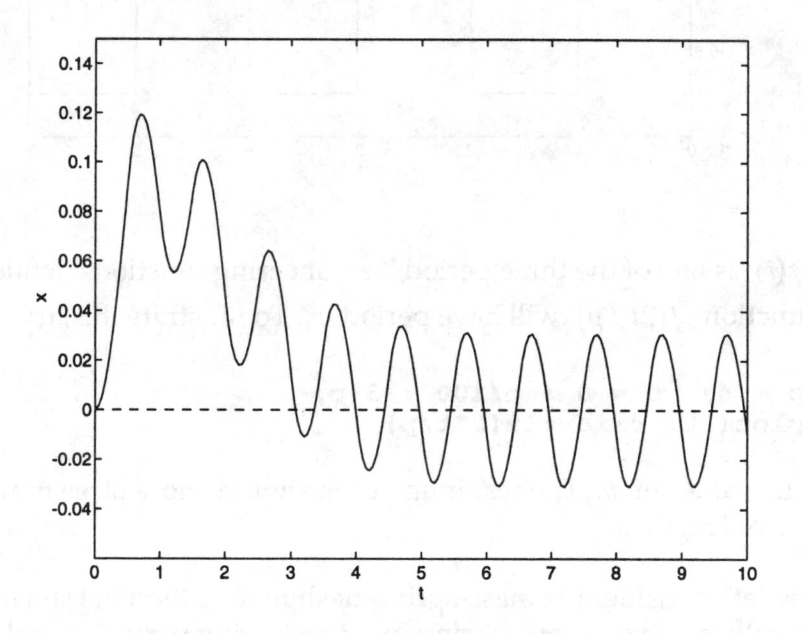

When we plot this solution, we see that after an initial transient dies out, the response function $x(t)$ settles down (as expected?) to a periodic oscillation with the same period as the input.

Chapter 8

Power Series Methods

Project 32
Computer Algebra Implementation of the Power Series Method

Reference: Section 8.2 of Edwards & Penney
DIFFERENTIAL EQUATIONS with Computing and Modeling

Here we illustrate the use of a computer algebra system to apply the power series method. As a first example, follow along with us in using power series to solve the simple differential equation

$$y'' + y = 0 \tag{1}$$

of Example 4 in Section 8.1 (whose familiar general solution is a linear combination of $\cos x$ and $\sin x$).

You can then assemble the sequence of commands we illustrate into a "template" for application to differential equations such as those in Problems 1 through 15 in Section 8.2.

An interesting application is to derive in this way some Legendre polynomials like those listed at the end of Section 8.2 in the text. For instance, try substituting a formal power series into the Legendre equation

$$(1 - x^2)\, y'' - 2x\, y' + 30\, y = 0$$

of order $n = 5$. One of the two linearly independent solutions you obtain should be a constant multiple of the 5th-degree Legendre polynomial $P_5(x)$. The Hermite polynomials of Problem 33 in Section 8.2 can be derived similarly.

Using *Maple*

First we write the initial 10 terms of a proposed series solution of our differential equation in (1).

```
> c := array(0..9):
> y := sum(c[n]*x^n, n = 0..9);
```

$$y := c_0 + c_1 x + c_2 x^2 + c_3 x^3 + c_4 x^4 + c_5 x^5 + c_6 x^6 + c_7 x^7 + c_8 x^8 + c_9 x^9$$

Then we substitute this series (partial sum) in the left-hand side of Equation (1) and collect coefficients of like powers of x,

```
> deq1 := diff(y, x$2) + y:
> deq1 := collect(deq1, x);
```

$$deq1 := c_9 x^9 + c_8 x^8 + (72\, c_9 + c_7)\, x^7 + (56\, c_8 + c_6)\, x^6 + (42\, c_7 + c_5)\, x^5 + (30\, c_6 + c_4)\, x^4 + (20\, c_5 + c_3)\, x^3 + (12\, c_4 + c_2)\, x^2 + (6\, c_3 + c_1)\, x + 2\, c_2 + c_0$$

Next we want to set up the equations the successive coefficients must satisfy. We do this by defining an array and then filling the elements of this array by equating (in turn) each of the series coefficients to zero.

```
> eq := array(0..7):
> for n from 0 to 7 do
        eq[n] := coeff(deq1,x,n) = 0:
        od:
> coeffEqs := convert(eq,set);
```

$$coeffEqs := \{2\, c_2 + c_0 = 0,\ 6\, c_3 + c_1 = 0,\ 12\, c_4 + c_2 = 0,\ 20\, c_5 + c_3 = 0,\ 30\, c_6 + c_4 = 0,\ 42\, c_7 + c_5 = 0,\ 56\, c_8 + c_6 = 0,\ 72\, c_9 + c_7 = 0\}$$

We have here a collection of eight linear equations relating the ten coefficients (**c[0]** through **c[9]**). We therefore should be able to solve for the successive coefficients

```
> succCoeffs := convert([seq(c[n], n=2..9)], set);
```

$$succCoeffs := \{c_2, c_3, c_4, c_5, c_6, c_7, c_8, c_9\}$$

in terms of **c[0]** and **c[1]**.

```
> ourCoeffs := solve(coeffEqs, succCoeffs);
```

$$ourCoeffs := \left\{c_4 = \frac{1}{24}\, c_0,\ c_2 = -\frac{1}{2}\, c_0,\ c_9 = \frac{1}{362880}\, c_1,\ c_8 = \frac{1}{40320}\, c_0,\ c_7 = -\frac{1}{5040}\, c_1,\ c_6 = -\frac{1}{720}\, c_0,\ c_5 = \frac{1}{120}\, c_1,\ c_3 = -\frac{1}{6}\, c_1\right\}$$

Now we substitute all these coefficients back into the original y-series.

```
> genSoln := subs(ourCoeffs, y);
```

$$genSoln := c_0 + c_1 x - \frac{1}{2}c_0 x^2 - \frac{1}{6}c_1 x^3 + \frac{1}{24}c_0 x^4 + \frac{1}{120}c_1 x^5 - \frac{1}{720}c_0 x^6 - \frac{1}{5040}c_1 x^7 + \frac{1}{40320}c_0 x^8 + \frac{1}{362880}c_1 x^9$$

We can finally extract our two linearly independent solutions by picking out the coefficients of `c[0]` and `c[1]`.

```
> genSoln := collect(genSoln, {c[0],c[1]}):
> ourCos := coeff(genSoln, c[0]);
```

$$ourCos := 1 - \frac{1}{2}x^2 + \frac{1}{24}x^4 - \frac{1}{720}x^6 + \frac{1}{40320}x^8$$

```
> ourSin := coeff(genSoln, c[1]);
```

$$ourSin := x - \frac{1}{6}x^3 + \frac{1}{120}x^5 - \frac{1}{5040}x^7 + \frac{1}{362880}x^9$$

Of course, we recognize here the initial terms of the familiar cosine and sine series, respectively.

Maple users can employ the built-in **dsolve** procedure with the **series** option to solve directly for power series solutions of a linear differential equation having $x = 0$ as an ordinary point. For example, with the Legendre equation

```
> deq2 :=
  (1-x^2)*diff(y(x),x$2)-2*x*diff(y(x),x)+30*y(x) = 0:
```

mentioned earlier, we obtain two (truncated) linearly independent series solutions by specifying the initial conditions as follows:

```
> Order := 12:
> dsolve(deq2, y(0)=0, D(y)(0)=1}, y(x), series);
```

$$y(x) = x - \frac{14}{3}x^3 + \frac{21}{5}x^5 + O(x^{12})$$

```
> dsolve({deq2, y(0)=1, D(y)(0)=0}, y(x), series);
```

$$y(x) = 1-15x^2+30x^4-10x^6-\frac{15}{7}x^8-x^{10}+O(x^{12})$$

The first of these provides a constant multiple of the standard 5th-degree Legendre polynomial $P_5(x)$.

Using *Mathematica*

First we write the initial ten terms of a proposed series solution:

```
y  =  Sum[ c[n] x^n, {n, 0, 9} ] + O[x]^10

                       2          3          4          5
c[0] + c[1] x + c[2] x  + c[3] x  + c[4] x  + c[5] x  +

       6          7          8          9      10
 c[6] x  + c[7] x  + c[8] x  + c[9] x  + O[x]
```

Then we substitute this series into the differential equation in (1),

```
deq1  =  D[ y, x, x ] + y  ==  0

                                                           2
(c[0] + 2 c[2]) + (c[1] + 6 c[3]) x + (c[2] + 12 c[4]) x  +

                   3                    4
 (c[3] + 20 c[5]) x  + (c[4] + 30 c[6]) x  +

                   5                    6
 (c[5] + 42 c[7]) x  + (c[6] + 56 c[8]) x  +

                   7       8
 (c[7] + 72 c[9]) x  + O[x]   == 0
```

Note that *Mathematica* has automatically collected coefficients of like powers for us! In fact, we can ask it to extract the equations the successive coefficients satisfy, using the command **LogicalExpand[series == 0]** that equates each of the series coefficients to zero.

```
coeffEqns  =  LogicalExpand[ deq1 ]

c[0] + 2 c[2] == 0 && c[1] + 6 c[3] == 0 &&

  c[2] + 12 c[4] == 0 && c[3] + 20 c[5] == 0 &&

  c[4] + 30 c[6] == 0 && c[5] + 42 c[7] == 0 &&

  c[6] + 56 c[8] == 0 && c[7] + 72 c[9] == 0
```

The double ampersand **&&** is just "and" in *Mathematica*, so we have here a collection of eight linear equations relating the ten coefficients (**c[0]** through **c[9]**). We therefore should be able to solve for the successive coefficients

```
succCoeffs  =  Table[ c[n], {n, 2, 9} ]

{c[2], c[3], c[4], c[5], c[6], c[7], c[8], c[9]}
```

in terms of **c[0]** and **c[1]**.

```
coeffs  =  Solve[ coeffEqns, succCoeffs ]

             c[0]             c[1]              -c[0]
{{c[8] -> -----, c[9] -> ------, c[2] -> -----,
             40320            362880              2

          -c[1]           c[0]           c[1]            -c[0]
  c[3] -> -----, c[4] -> ----, c[5] -> ----, c[6] -> -----,
            6              24             120              720

          -c[1]
  c[7] -> -----}}
           5040
```

Now we substitute all these coefficients back into the series.

```
genSoln  =  y /. coeffs

                              2          3          4          5
                       c[0] x     c[1] x     c[0] x     c[1] x
{c[0] + c[1] x - ------- - ------- + ------- + ------- -
                          2          6          24         120

          6          7          8          9
    c[0] x     c[1] x     c[0] x     c[1] x          10
    ------- - ------- + ------- + ------- + O[x]   }
      720       5040      40320     362880
```

We can finally extract our two linearly independent solutions by picking out the coefficients of **c[0]** and **c[1]**.

```
ourCos  =  Coefficient[ genSoln, c[0] ]

        2    4     6      8
       x    x     x      x
{1 - -- + -- - --- + -----}
       2    24   720    40320

ourSin  =  Coefficient[ genSoln, c[1] ]
```

```
       3     5      7       9
      x     x      x       x
{x - -- + --- - ---- + ------}
      6   120   5040   362880
```

Of course, we recognize here the initial terms of the cosine and sine series, respectively.

Using MATLAB

First we write the initial eight terms of a proposed series solution of our differential equation.

```
» y ='c0+c1*x+c2*x^2+c3*x^3+c4*x^4+c5*x^5+c6*x^6+c7*x^7';
» pretty(y)

                   2        3        4        5        6        7
c0 + c1 x + c2 x  + c3 x  + c4 x  + c5 x  + c6 x  + c7 x
```

Then we substitute this series (partial sum) in the left-hand side of Equation (1) and collect coefficients of like powers of x,

```
» ddy = diff(y,'x',2);          % the 2nd derivative of y
» deq1 = symop(ddy,'+',y);      % the diff. equation
» deq1 = collect(deq1)

deq1 =
c7*x^7+c6*x^6+(42*c7+c5)*x^5+(30*c6+c4)*x^4+
(c3+20*c5)*x^3+(c2+12*c4)*x^2+(6*c3+c1)*x+2*c2+c0
```

Next we want to set up the equations the original coefficients in **y** must satisfy. By cutting and pasting it is a simple matter to pick out the successive coefficients in **deq1** and equate each of them to zero..

```
» eq1 = '2*c2+c0=0';      eq2 = '6*c3+c1=0';
» eq3 = 'c2+12*c4=0';     eq4 = 'c3+20*c5=0';
» eq5 = '30*c6+c4=0';     eq6 = '42*c7+c5=0';
```

We have here a collection of six linear equations relating the eight coefficients (**c0** through **c7**). We therefore should be able to solve for the successive coefficients **c2** through **c7** in terms of c[0] and c[1].

```
» [c2,c3,c4,c5,c6,c7] =
  solve(eq1,eq2,eq3,eq4,eq5,eq6,'c2,c3,c4,c5,c6,c7');

» pretty
```

```
{c2 = - 1/2 c0, c3 = - 1/6 c1, c4 = 1/24 c0,
 c6 = - 1/720 c0, c5 = 1/120 c1, c7 = - 1/5040 c1}
```

Before substitut'ng these coefficients back into the original y-series, let's set the arbitrary coefficients `c0` and `c1` equal to 1.

```
» c0 = 1;   c2 = subs(c2,1,'c0');   c4 = subs(c4,1,'c0')
» c1 = 1;   c3 = subs(c3,1,'c1');   c5 = subs(c5,1,'c1');
```

When we collect the terms of even degree,

```
» ourCos = symop(c0,'+',c2,'*x^2','+',c4,'*x^4');
» pretty

                              2          4
                   1 - 1/2 x  + 1/24 x
```

and the terms of odd degree,

```
» ourSin = symop(c1,'+',c3,'*x^3','+',c5,'*x^5');
» pretty

                              3           5
                   1 - 1/6 x  + 1/120 x
```

we recognize the initial terms of the familiar cosine and sine series, respectively.

Project 33

Computer Algebra Implementation of the Frobenius Series Method

Reference: Section 8.3 of Edwards & Penney
DIFFERENTIAL EQUATIONS with Computing and Modeling

Here we illustrate the use of a computer algebra system to apply the method of Frobenius. In the paragraphs that follow, we consider the differential equation

$$2x^2 y'' + 3x\, y' - (x^2 + 1)y = 0 \qquad (1)$$

of Example 4 in Section 8.3 of the text, where we found the two indicial roots $r_1 = \frac{1}{2}$ and $r_2 = -1$. We carry through the formal Frobenius method starting with the larger indicial root r_1, and you can then apply the same process to derive the second Frobenius series solution (found manually in the text) corresponding to r_2.

In the following examples, use this method to derive Frobenius series solutions that can be checked against the given known general solutions.

1. $x\,y'' - y' + 4x^3\,y = 0, \qquad y(x) = A\cos x^2 + B\sin x^2$

2. $x\,y'' - 2\,y' + 9x^5\,y = 0, \qquad y(x) = A\cos x^3 + B\sin x^3$

3 $4x\,y'' + 2\,y' + y = 0, \qquad y(x) = A\cos\sqrt{x} + B\sin\sqrt{x}$

4. $x\,y'' + 2\,y' + x\,y = 0, \qquad y(x) = \frac{1}{x}(A\cos x + B\sin x)$

5. $4x\,y'' + 6\,y' + y = 0, \qquad y(x) = \frac{1}{\sqrt{x}}\left(A\cos\sqrt{x} + B\sin\sqrt{x}\right)$

6. $x\,y'' + x\,y' + (4x^4 - 1)\,y = 0, \qquad y(x) = \frac{1}{x}\left(A\cos x^2 + B\sin x^2\right)$

7. $x\,y'' + 3\,y' + 4x^3\,y = 0, \qquad y(x) = \frac{1}{x^2}\left(A\cos x^2 + B\sin x^2\right)$

8. $x^2\,y'' + x^2\,y' - 2\,y = 0, \qquad y(x) = \frac{1}{x}\left[A\,(2-x) + B\,(2+x)\,e^{-x}\right]$

The following three problems involve the arctangent series

$$\tan^{-1} x = x - \frac{x^3}{3} + \frac{x^5}{5} - \frac{x^7}{7} + \cdots .$$

9. $(x + x^3)\,y'' + (2 + 4x^2)\,y' - 2x\,y = 0, \qquad y(x) = \frac{1}{x}\left(A + B\tan^{-1} x\right)$

10. $(2x + 2x^2)\,y'' + (3 + 5x)\,y' + y = 0, \qquad y(x) = \frac{1}{\sqrt{x}}\left(A + B\tan^{-1}\sqrt{x}\right)$

11. $(x + x^5)\,y'' + (3 + 7x^4)\,y' + 8x^3\,y = 0, \qquad y(x) = \frac{1}{x^2}\left(A + B\tan^{-1} x^2\right)$

Using *Maple*

Beginning with the indicial root $r_1 = \frac{1}{2}$, we first write the initial seven terms of a proposed Frobenius series solution:

```
> a := array(0..6):
> y := x^(1/2)*sum( a[n]*x^(n), n = 0..6);
```

$$y := \sqrt{x}\left(a_0 + a_1 x + a_2 x^2 + a_3 x^3 + a_4 x^4 + a_5 x^5 + a_6 x^6\right)$$

Then we substitute this series (partial sum) into the left-hand side of Equation (1),

```
> deq1 := 2*x^2*diff(y,x$2)+3*x*diff(y,x)-(x^2+1)*y:
> deq1 := simplify(deq1);
```

$$deq1 := -x^{3/2}\left(-5a_1 - 14a_2x - 27a_3x^2 - 44a_4x^3 - 65a_5x^4 - 90a_6x^5 + a_0x + x^6a_5 + x^7a_6 + x^2a_1 + x^3a_2 + x^4a_3 + x^5a_4\right)$$

Noting the $x^{3/2}$ factor, we multiply by $x^{-3/2}$ and then collect coefficients of like powers of x.

```
> deq2 := collect( x^(-3/2)*deq1, x);
```

$$deq2 := -x^7a_6 - x^6a_5 + (90a_6 - a_4)x^5 + (-a_3 + 65a_5)x^4 + (-a_2 + 44a_4)x^3 + (-a_1 + 27a_3)x^2 + (14a_2 - a_0)x + 5a_1$$

Next we set up the equations the successive coefficients must satisfy. We do this by defining an array and then filling the elements of this array by equating (in turn) each of the series coefficients to zero.

```
> eqs := array(0..5):
> for n from 0 to 5 do
      eqs[n] := coeff(deq1,x,n) = 0:
      od:
> coeffEqs := convert(eqs, set);
```

$$coeffEqs := \{5\,a_1 = 0,\ -a_2 + 44\,a_4 = 0,\ -a_3 + 65\,a_5 = 0,\ 90\,a_6 - a_4 = 0,\ 14\,a_2 - a_0 = 0,\ -a_1 + 27\,a_3 = 0\}$$

We have here a collection of six linear equations relating the seven coefficients (`a[0]` through `a[6]`). Hence we should be able to solve for the successive coefficients

```
> succCoeffs := convert([seq(a[n], n=1..6)], set);
```

$$succCoeffs := \{a_2, a_3, a_4, a_5, a_6, a_1\}$$

in terms of `a[0]`.

```
> ourCoeffs := solve(coeffEqs, succCoeffs);
```

$$ourCoeffs := \left\{a_1 = 0, a_6 = \frac{1}{55440}a_0, a_4 = \frac{1}{616}a_0, a_2 = \frac{1}{14}a_0, a_5 = 0, a_3 = 0\right\}$$

Finally we substitute all these coefficients back into the original series.

```
> partSoln := subs(ourCoeffs, y);
```

$$partSoln := \sqrt{x}\left(a_0 + \frac{1}{14}a_0\, x^2 + \frac{1}{616}a_0\, x^4 + \frac{1}{55440}a_0\, x^6\right)$$

Note that (after factoring out a_0) this result agrees with the first particular solution

$$y_1(x) \;=\; a_0 x^{1/2}\left(1+\frac{x^2}{14}+\frac{x^4}{616}+\frac{x^6}{55440}+\cdots\right).$$

found in the text.

Using *Mathematica*

Beginning with the indicial root $r_1 = \frac{1}{2}$, we first write the initial seven terms of a proposed Frobenius series solution:

```
y = x^(1/2) Sum[ a[n] x^n, {n, 0, 6} ] + O[x]^7

                          3/2           5/2           7/2
a[0] Sqrt[x] + a[1] x        + a[2] x      + a[3] x       +

         9/2          11/2          13/2       7
  a[4] x     + a[5] x      + a[6] x     + O[x]
```

Then we substitute this series into the differential equation in (1),

```
deq1 = 2x^2 D[y, x,x] + 3x D[y, x] - (x^2 + 1)y == 0
```

```
                  3/2                       5/2
5 a[1] x     + (-a[0] + 14 a[2]) x     +

                            7/2                      9/2
   (-a[1] + 27 a[3]) x     + (-a[2] + 44 a[4]) x     +

                            11/2                      13/2
   (-a[3] + 65 a[5]) x      + (-a[4] + 90 a[6]) x      +

       7
   O[x]   == 0
```

Mathematica has automatically collected like powers for us, and we can use the **LogicalExpand** command to extract the equations that the successive coefficients satisfy.

```
coeffEqns  =  LogicalExpand[ deq1 ]

5 a[1] == 0 && -a[0] + 14 a[2] == 0 &&

  -a[1] + 27 a[3] == 0 && -a[2] + 44 a[4] == 0 &&

  -a[3] + 65 a[5] == 0 && -a[4] + 90 a[6] == 0
```

We have here a collection of six linear equations relating the seven coefficients (**a[0]** through **a[6]**). Hence we should be able to solve for the successive coefficients

```
succCoeffs  =  Table[ a[n], {n, 1, 6} ]

{a[1], a[2], a[3], a[4], a[5], a[6]}
```

in terms of **a[0]**.

```
ourCoeffs  =  Solve[ coeffEqns, succCoeffs ]

                         a[0]                        a[0]
{{a[5] -> 0, a[6] -> -----, a[1] -> 0, a[2] -> ----,
                         55440                       14

                         a[0]
   a[3] -> 0, a[4] -> ----}}
                         616
```

Finally we substitute all these coefficients back into the original series.

```
partSoln  =  y  /.  ourCoeffs
```

```
                       5/2            9/2            13/2
                a[0] x         a[0] x         a[0] x
{a[0] Sqrt[x] + --------- + --------- + ---------- +
                   14             616           55440

      7
   O[x]  }
```

Note that (after factoring out $a_0x^{1/2}$) this result agrees with the first particular solution

$$y_1(x) = a_0x^{1/2}\left(1+\frac{x^2}{14}+\frac{x^4}{616}+\frac{x^6}{55440}+\cdots\right).$$

found in the text.

Using MATLAB

Beginning with the indicial root $r_1 = \frac{1}{2}$, we first write the initial seven terms of a proposed Frobenius series solution:

```
» y =
  'x^(1/2)*(a0+a1*x+a2*x^2+a3*x^3+a4*x^4+a5*x^5+a6*x^6)';
» pretty(y)

 1/2                     2         3         4         5         6
x    (a0 + a1 x + a2 x  + a3 x  + a4 x  + a5 x  + a6 x )
```

Then we substitute this series (partial sum) into the left-hand side of Equation (1), which we assemble term by term:

```
» Dy = diff(y,'x');
» D2y = diff(Dy,'x');

» term1 = symmul('2*x^2',D2y);
» term2 = symmul('3*x',Dy);
» term3 = symmul('(x^2+1)',y);

» deq1 = symop(term1,'+',term2,'-',term3);
» simple(deq1)

ans =
-x^(3/2)*
(-5*a1-14*a2*x-27*a3*x^2-44*a4*x^3-65*a5*x^4-90*a6*x^5+
x*a0+x^2*a1+x^3*a2+x^4*a3+x^5*a4+x^6*a5+x^7*a6)
```

Noting the $x^{3/2}$ factor, we multiply by $x^{-3/2}$ and then collect coefficients of like powers of x.

```
» deq2 = collect(simplify(symmul('x^(-3/2)',deq1)))
deq2 =
-x^7*a6-x^6*a5+(90*a6-a4)*x^5+(-a3+65*a5)*x^4+
(-a2+44*a4)*x^3+(-a1+27*a3)*x^2+(14*a2-a0)*x+5*a1
```

Next we set up the equations the original coefficients in **y** must satisfy. By cutting and pasting it is a simple matter to pick out the successive coefficients in **deq2** and equate each of them to zero.

```
» eq1 = '5*a1 = 0';              eq2 = '14*a2-a0 = 0';
» eq3 = '-a1+27*a3 = 0';         eq4 = '-a2+44*a4 = 0';
» eq5 = '-a3+65*a5 = 0';         eq6 = '90*a6-a4 = 0';
```

We have here a collection of six linear equations relating the seven coefficients (**a[0]** through **a[6]**). Hence we should be able to solve for the successive coefficients **a1** through **a6** in terms of **a0**.

```
» [a1,a2,a3,a4,a5,a6] =
solve(eq1,eq2,eq3,eq4,eq5,eq6,'a1,a2,a3,a4,a5,a6');
» pretty

          {a6 = 1/55440 a0, a1 = 0, a2 = 1/14 a0,
                a4 = 1/616 a0, a5 = 0, a3 = 0}
```

Finally we substitute all these coefficients back into the original series.

```
» y = subs(y,a1,'a1');  y = subs(y,a2,'a2');
» y = subs(y,a3,'a3');  y = subs(y,a4,'a4');
» y = subs(y,a5,'a5');  y = subs(y,a6,'a6');
» pretty

  1/2                  2              4                   6
 x     (a0 + 1/14 a0 x  + 1/616 a0 x  + 1/55440 a0 x )
```

Note that (after factoring out a_0) this result agrees with the first particular solution

$$y_1(x) \;=\; a_0 x^{1/2}\left(1+\frac{x^2}{14}+\frac{x^4}{616}+\frac{x^6}{55440}+\cdots\right).$$

found in the text.

Project 34

The Exceptional Case: Computer Algebra Implementation of Reduction of Order

Reference: Section 8.4 of Edwards & Penney *DIFFERENTIAL EQUATIONS with Computing and Modeling*

The reduction of order formula in Eq. (28) of this section is readily implemented using a computer algebra system. Given a known solution $y_1(x)$ of the homogeneous linear second-order equation

$$y'' + P(x)\,y' + Q(x)\,y \;=\; 0 \tag{1}$$

on an open interval where P and Q are continuous, the **reduction of order formula** gives a second independent solution $y_2(x)$ defined by

$$y_2(x) \;=\; y_1(x) \int \frac{\exp(-\int P(x)\,dx)}{y_1(x)^2}\,dx. \tag{2}$$

In the paragraphs below we illustrate the use of (2) with *Maple*, *Mathematica*, and MATLAB to derive a second solution of Bessel's equation of order zero, beginning with the known power series solution

$$J_0(x) \;=\; \sum_{n=0}^{\infty} (-1)^n \frac{x^{2n}}{2^{2n}\,(n!)^2}. \tag{3}$$

After verifying (with your computer algebra system) the computations we present, you can begin with the power series for $J_1(x)$ in Eq. (49) of Section 8.4 and derive similarly the second solution in (50) of Bessel's equation of order 1. Problems 9 through 14 in Section 8.4 can also be partially automated in this way.

Using *Maple*

We begin by entering the initial terms of the the known solution (3) in the form

```
> y1 := sum( (-x^2/4)^n / (n!)^2, n = 0..4 );
```

$$y1 := 1 - \frac{1}{4}x^2 + \frac{1}{64}x^4 - \frac{1}{2304}x^6 + \frac{1}{147456}x^8$$

Then with

```
> P := 1/x:
```

we need only "build" the integral of Eq. (2) and evaluate it:

```
> part1 := exp( -int(P,x) );
```

$$part1 := \frac{1}{x}$$

```
> part2 :=series(1/y1^2,x,9):
  part2 := convert(part2, polynom);
```

$$part2 := 1+\frac{1}{2}x^2+\frac{5}{32}x^4+\frac{23}{5764}x^6+\frac{677}{73728}x^8$$

```
> integral := int( part1*part2, x );
```

$$integral := \frac{677}{589824}x^8+\frac{23}{3456}x^6+\frac{5}{128}x^4+\frac{1}{4}x^2+\ln(x)$$

Then the computation

```
> y := series(y1*(integral - ln(x)),x,9);
```

$$y := \frac{1}{4}x^2-\frac{3}{128}x^4+\frac{11}{13824}x^6-\frac{25}{1769472}x^8+O(x^{10})$$

yields the second solution

```
> y2 := J0*ln(x) + y;
```

$$y2 := J0\ln(x)+\frac{1}{4}x^2-\frac{3}{128}x^4+\frac{11}{13824}x^6-\frac{25}{1769472}x^8+O(x^{10})$$

of Bessel's equation of order 0 (as we see it in Eq. (45) of the text).

Using *Mathematica*

We begin by entering the known solution (3) in the form

```
y1  =  Sum[ (-x^2/4)^n /(n!)^2, {n, 0, 4} ] + O[x]^9
```

```
       2     4      6        8
      x     x      x        x          9
1 -  -- +  -- -  ---- +  ------ + O[x]
      4    64    2304    147456
```

Then with

```
P  =  1/x;
```

we need only substitute in the integral of Eq. (2):

```
integral  =  Integrate[ Exp[ -Integrate[P,x]]/y1^2, x ]
```

```
              2       4       6       8
             x     5 x     23 x    677 x     9
Log[x]  +   -- +  ---- +  ----- + ------ + O[x]
             4     128     3456   589824
```

Then the computation

```
y  =  y1 (integral - Log[x])
```

```
 2      4       6          8
x    3 x     11 x      25 x        9
-- - ---- + ----- - ------- + O[x]
4    128    13824   1769472
```

yields the second solution

```
y2  =  J0 Log[x] + y
```

```
                  2      4        6          8
                 x    3 x     11 x      25 x        9
J0 Log[x] +     -- - ---- +  ----- - ------- + O[x]
                 4    128    13824   1769472
```

of Bessel's equation of order 0 (as we see it in Eq. (45) of the text).

Using MATLAB

We begin by entering the initial terms of the the known solution (3) in the form

```
» y1 = '1 - x^2/4 + x^4/64 - x^6/2304 + x^8/147456';
» pretty(y1)

                  2          4            6              8
      1 - 1/4 x  + 1/64 x  - 1/2304 x  + 1/147456 x
```

Then with

```
» P = '1/x';   int(P)
ans =
log(x)
```

we need only "build" the integral of Eq. (2) and evaluate it. Using the **taylor** function to compute series expansions, we calculate

```
» part1 = simplify('exp(-log(x))')
part1 =
1/x

» part2 = taylor(sympow(y1,-2),10);
» part2 = symsub(part2,'O(x^10)');
» pretty(part2)

                     2          4    23  6    677   8
      1 + 1/2 x  + 5/32 x  + --- x  + ----- x
                                    576       73728

» integral = int( symmul(part1,part2) );
» pretty(integral)

      677    8     23   6           4          2
     ------ x  + ---- x  + 5/128 x  + 1/4 x  + log(x)
     589824       3456
```

Then the computation

```
» y = taylor( symop(y1,'*',symsub(integral,'log(x)') ),9);
» pretty(y)

         2            4      11    6       25       8        10
   1/4 x  - 3/128 x  + ----- x  - ------- x  + O(x   )
                        13824      1769472
```

yields the second solution

```
» y2 = symadd('J0*log(x)',y);
» pretty(y2)

                     2            4      11    6       25       8        10
 J0 log(x) + 1/4 x  - 3/128 x  + ----- x  - ------- x  + O(x   )
                                  13824      1769472
```

of Bessel's equation of order 0 (as we see it in Eq. (45) of the text).

Project 35

Application of Bessel Functions to Riccati Equations

Reference: Section 8.6 of Edwards & Penney
DIFFERENTIAL EQUATIONS with Computing and Modeling

A Riccati equation is one of the form

$$y' = A(x)\,y^2 + B(x)\,y + C(x).$$

Many Riccati equations like the ones listed below can be solved explicitly in terms of Bessel functions.

$$y' = x^2 + y^2 \tag{1}$$

$$y' = x^2 - y^2 \tag{2}$$

$$y' = y^2 - x^2 \tag{3}$$

$$y' = x + y^2 \tag{4}$$

$$y' = x - y^2 \tag{5}$$

$$y' = y^2 - x \tag{6}$$

For instance, Problem 15 in Section 8.6 of the text says that the general solution of (1) is given by

$$y(x) = x\frac{J_{3/4}\left(\frac{1}{2}x^2\right) - cJ_{-3/4}\left(\frac{1}{2}x^2\right)}{cJ_{1/4}\left(\frac{1}{2}x^2\right) + J_{-1/4}\left(\frac{1}{2}x^2\right)}. \tag{7}$$

See whether the symbolic DE solver command in your computer algebra system, such as

```
dsolve(diff(y(x),x) = x^2 + y(x)^2, y(x))
```

(*Maple*)

or

```
DSolve[y'[x] == x^2 + y[x]^2, y[x], x]
```

(*Mathematica*)

or

```
» dsolve('Dy = x^2 + y^2')
```

(MATLAB)

agrees with (7). (*Note:* Using *Maple* or a *Maple*-based MATLAB system, you will need *Maple* V release 3 or later. Using*Mathematica* version 2.1 or later you may need to load the **Calculus`DSolve`** package before executing the **DSolve** command.)

If Bessel functions other than those appearing in (7) are involved, you may need to apply Identities (26) and (27) in Section 8.5 of the text to transform the computer's "answer" to (7). Then see whether your system can take the limit as $x \to 0$ in (7) to show that the arbitrary constant c is given in terms of the initial value $y(0)$ by

$$c = -\frac{y(0)\,\Gamma\left(\frac{1}{4}\right)}{2\,\Gamma\left(\frac{3}{4}\right)} \tag{8}$$

Now you should be able to use built-in Bessel functions to plot typical solution curves like those illustrated below.

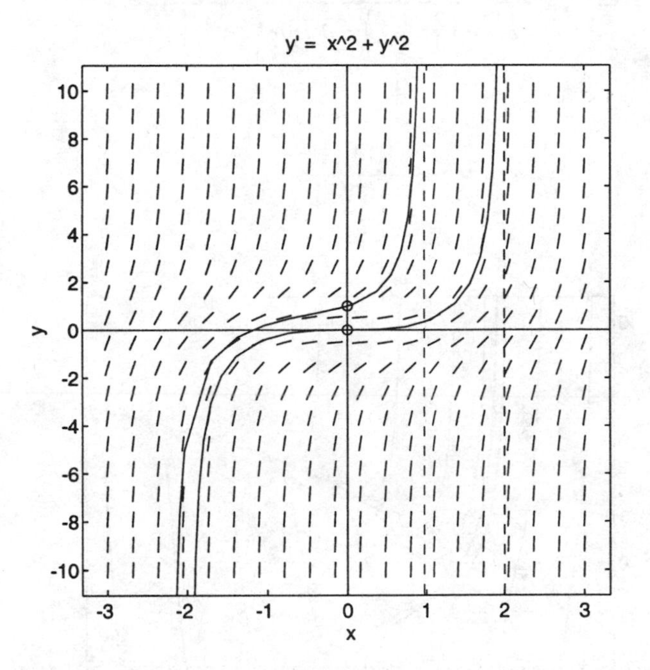

This figure shows the trajectories of (1) satisfying the initial condtions $y(0) = 0$ and $y(0) = 1$, together with apparent vertical asymptotes. Can you use (7) and (8) to verify that these asymptotes are given approximately by $x = 2.00315$ and $x = 0.96981$, respectively? (*Suggestion*: You are looking for zeros of the denominator in (7).)

Next, investigate similarly one of the other equations in (2)–(6). Each has a general solution of the same general form as in (7) — a quotient of linear combinations of Bessel functions. In addition to $J_p(x)$ and $Y_p(x)$, these solutions may involve the *modified Bessel functions*

$$I_p(x) = i^{-p} J_p(ix)$$

and

$$K_p(x) = \frac{\pi}{2} i^{-p} \left[J_p(ix) + Y_p(ix) \right]$$

that satisfy the *modified Bessel equation*

$$x^2 y'' + x\, y' - (x^2 + p^2) y \;=\; 0$$

of order p. For instance, the general solution of Equation (5) is given for $x > 0$ by

$$y(x) \;=\; \sqrt{x}\, \frac{I_{2/3}\left(\frac{2}{3}x^{3/2}\right) - c\, I_{-2/3}\left(\frac{2}{3}x^{3/2}\right)}{I_{-1/3}\left(\frac{2}{3}x^{3/2}\right) - c\, I_{1/3}\left(\frac{2}{3}x^{3/2}\right)} \tag{9}$$

where

$$c \;=\; -\frac{y(0)\, \Gamma\left(\frac{1}{3}\right)}{\sqrt[3]{3}\, \Gamma\left(\frac{2}{3}\right)}. \tag{10}$$

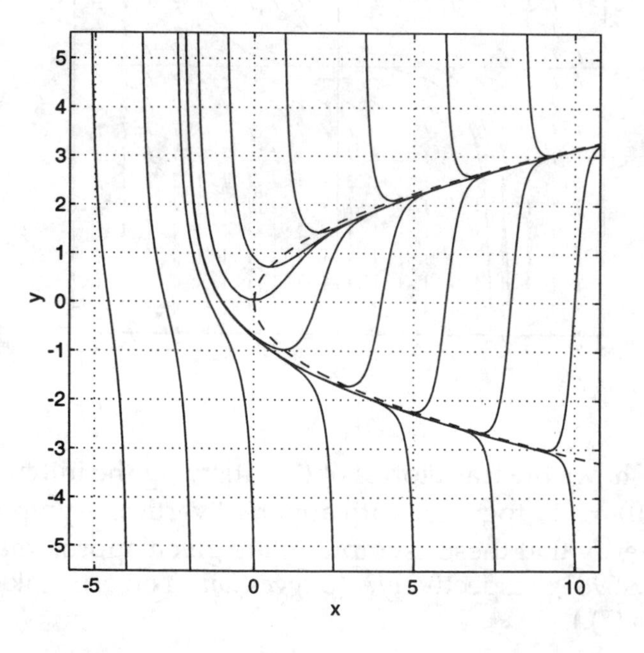

The preceding figure shows some typical solution curves, together with the parabola $y^2 = x$ that appears to bear an interesting relation to Equation (5) — we see a funnel near $y = +\sqrt{x}$ and a spout near $y = -\sqrt{x}$.

The Bessel functions with imaginary argument that appear in the definitions of $I_p(x)$ and $K_p(x)$ may look exotic, but the power series of the modified function $I_n(x)$ is simply that of the unmodified function $J_n(x)$, except without the alternating minus signs. For instance,

$$I_0(x) = 1 + \frac{x^2}{4} + \frac{x^4}{64} + \frac{x^6}{2304} + \cdots\cdots$$

and

$$I_1(x) = \frac{x}{2} + \frac{x^3}{16} + \frac{x^5}{384} + \frac{x^7}{18432} + \cdots\cdots .$$

Check these power series expansions using your computer algebra system — look at **BesselI** in either *Maple* or *Mathematica* — and compare them with Eqs. (17)–(18) in Section 8.5 of the text.

Chapter 9

Fourier Series Methods

Project 36
Computer Algebra Derivation of Fourier Series

Reference: Section 9.2 of Edwards & Penney *DIFFERENTIAL EQUATIONS with Computing and Modeling*

A computer algebra system can greatly ease the burden of calculation of the Fourier coefficients of a given function $f(t)$. In the case of a function defined "piecewise," we must take care to "split" the integral according to the different intervals of definition of the function. In the paragraphs that follow we illustrate the use of *Maple*, *Mathematica*, and MATLAB in deriving the Fourier series

$$f(t) = \frac{4}{\pi} \sum_{n \text{ odd}}^{\infty} \frac{\sin nt}{n} \tag{1}$$

of the period 2π square wave function defined on $(-\pi, \pi)$ by

$$f(t) = \begin{cases} -1 & \text{if } -\pi < t < 0, \\ +1 & \text{if } 0 < t < \pi. \end{cases} \tag{2}$$

In this case the function is defined by different formulas on two different intervals, so each Fourier coefficient integral from $-\pi$ to π must be calculated as the sum of two integrals:

$$\begin{aligned} a_n &= \frac{1}{\pi}\int_{-\pi}^{0}(-1)\cos nt\,dt + \frac{1}{\pi}\int_{0}^{\pi}(+1)\cos nt\,dt, \\ b_n &= \frac{1}{\pi}\int_{-\pi}^{0}(-1)\sin nt\,dt + \frac{1}{\pi}\int_{0}^{\pi}(+1)\sin nt\,dt. \end{aligned} \tag{3}$$

To practice the symbolic derivation of Fourier series in this manner, you can begin by verifying the Fourier series calculated manually in Examples 1 and 2 of Section 9.2 in the text. Then Problems 1 through 21 there are fair game. Finally, the period 2π triangular wave and trapezoidal wave functions illustrated in the figures below have especially interesting Fourier series that we invite you to discover for yourself.

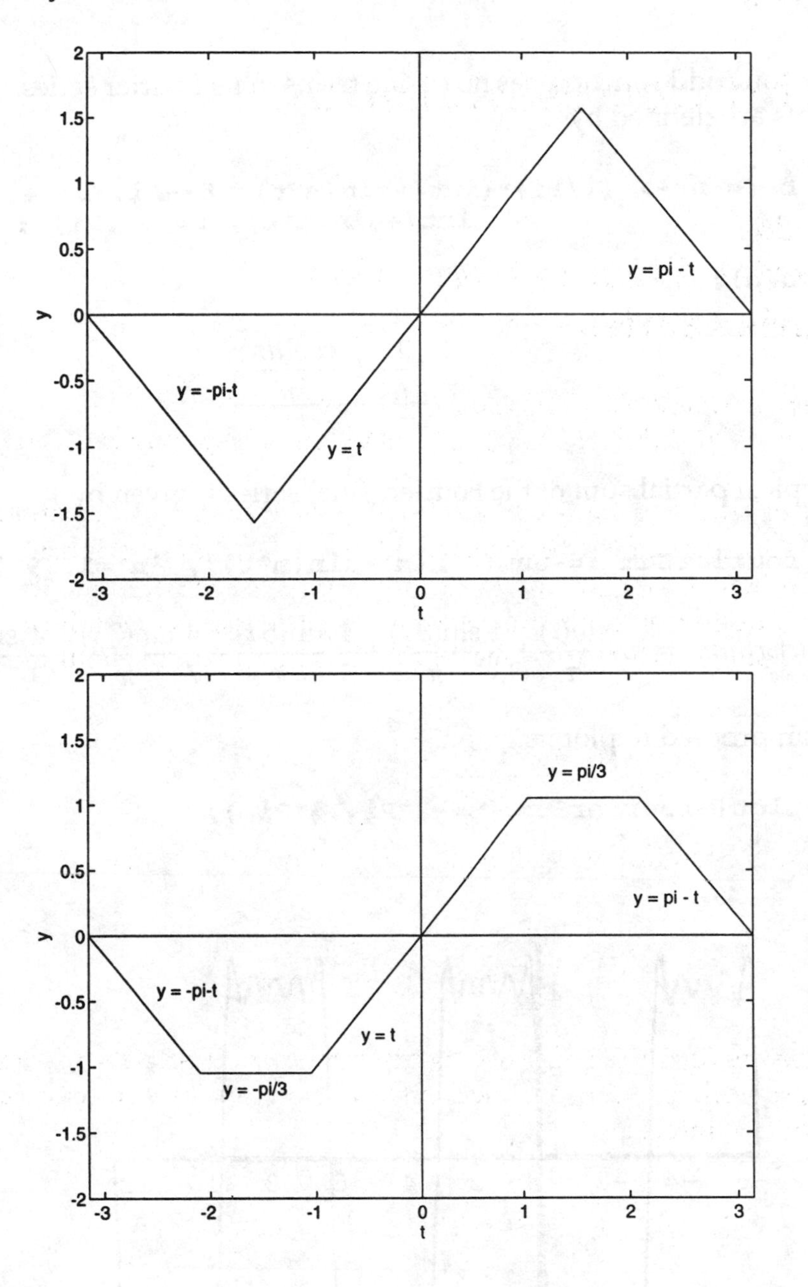

Using *Maple*

We can define the cosine coefficients in (3) as functions of n by

```
> a := n -> (1/Pi)*(int(-cos(n*t), t=-Pi..0) +
                     int(+cos(n*t), t=0..Pi)):
```

```
> a(n);
```

$$0$$

Of course, our odd function has no cosine terms in its Fourier series. The sine coefficients are defined by

```
> b := n -> (1/Pi)*(int(-sin(n*t), t=-Pi..0) +
                    int(+sin(n*t), t= 0..Pi)):
```

```
> b(n);
```

$$\frac{2\frac{1}{n} - 2\frac{\cos(n\pi)}{n}}{\pi}$$

Then a typical partial sum of the Fourier (sine) series is given by

```
> fourierSum := sum( 'b(n)*sin(n*t)', 'n'=1..9 );
```

$$fourierSum := 4\frac{\sin(t)}{\pi} + \frac{4}{3}\frac{\sin(3\,t)}{\pi} + \frac{4}{5}\frac{\sin(5\,t)}{\pi} + \frac{4}{7}\frac{\sin(7\,t)}{\pi} + \frac{4}{9}\frac{\sin(9\,t)}{\pi}$$

and we can proceed to plot its graph.

```
> plot( fourierSum, t=-2*Pi..4*Pi );
```

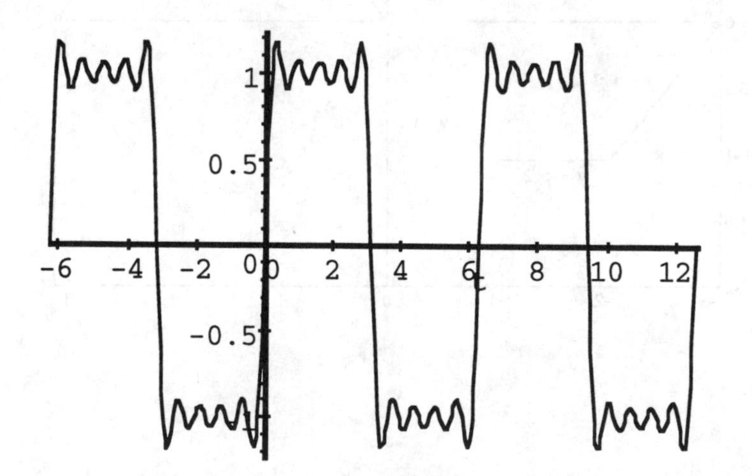

Using *Mathematica*

We can define the cosine coefficients in (3) as functions of n by

```
a[n_] = (1/Pi)*(Integrate[-Cos[n*t], {t,-Pi, 0} ] +
                Integrate[+Cos[n*t], {t, 0, Pi} ] )

0
```

Of course, our odd function has no cosine terms in its Fourier series. The sine coefficients are defined by

```
b[n_] = (1/Pi)*(Integrate[-Sin[n*t], {t,-Pi, 0}] +
                Integrate[+Sin[n*t], {t, 0, +Pi}])

2   2 Cos[n Pi]
- - -----------
n        n
---------------
      Pi
```

Then a typical partial sum of the Fourier (sine) series is given by

```
fourierSum = Sum[ b[n] Sin[n t], {n, 1, 9} ]

4 Sin[t]    4 Sin[3 t]   4 Sin[5 t]   4 Sin[7 t]   4 Sin[9 t]
-------- + ---------- + ---------- + ---------- + ----------
   Pi         3 Pi         5 Pi         7 Pi         9 Pi
```

and we can proceed to plot its graph.

```
Plot[ fourierSum, {t, -2 Pi, 4 Pi} ];
```

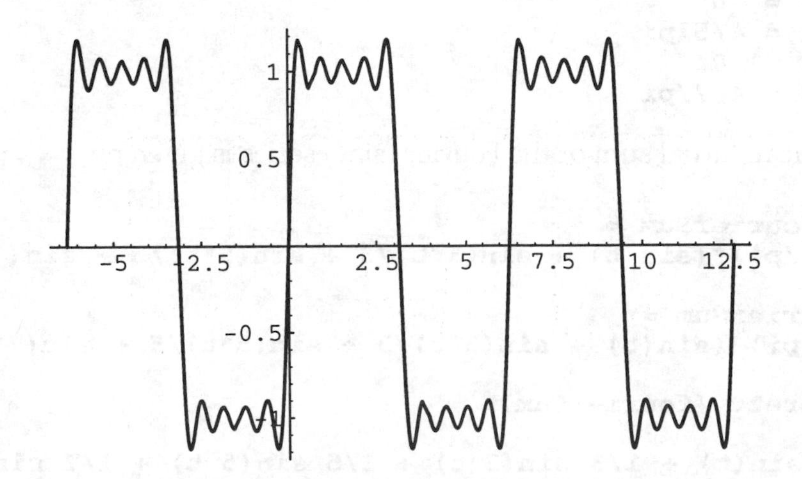

We can define the cosine coefficients in (3) as functions of n by

```
» an = symmul('1/pi',symadd(int('-cos(n*t)', -pi,0),
                            int( 'cos(n*t)',  0, pi) ))
an = 0
```

Of course, our odd function has no cosine terms in its Fourier series. The sine coefficients are defined by

```
» bn = symmul('1/pi',symadd(int('-sin(n*t)', -pi,0),
                            int( 'sin(n*t)',  0, pi) ))

bn = 1/pi*(2/n-2*cos(n*pi)/n)

» bn = subs(bn, '(-1)^n', 'cos(n*pi)');
» pretty(bn)
                                  n
                              (-1)
                   2/n - 2 -----
                                n
                   -------------
                         pi
```

The numerical values of the first seven coefficients are then given by

```
» for n = 1 : 7
     simplify( subs(bn,n,'n') )
     end

ans = 4/pi
ans =  0
ans = 4/3/pi
ans =  0
ans = 4/5/pi
ans =  0
ans = 4/7/pi
```

Thus a typical partial sum of the Fourier (sine) series is given by

```
» fourierSum =
'(4/pi)*(sin(t) + sin(3*t)/3 + sin(5*t)/5 + sin(7*t)/7)'

fourierSum =
(4/pi)*(sin(t) + sin(3*t)/3 + sin(5*t)/5 + sin(7*t)/7)

» pretty(fourierSum)

   sin(t) + 1/3 sin(3 t) + 1/5 sin(5 t) + 1/7 sin(7 t)
 4 ---------------------------------------------------
                           pi
```

and we can proceed to plot its graph.

```
» ezplot(fourierSum, [-2*pi 4*pi])
```

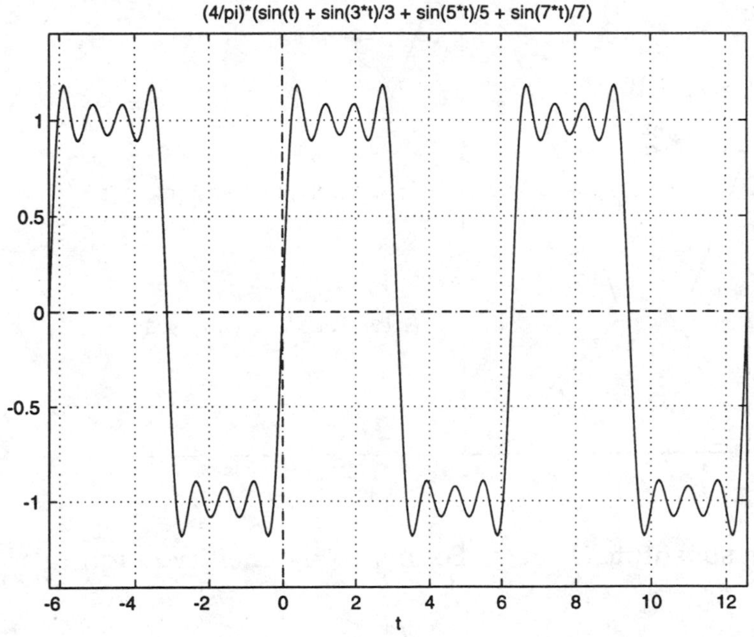

Project 37
More Computer Algebra Derivation of Fourier Series of Piecewise Smooth Functions

Reference: Section 9.3 of Edwards & Penney
DIFFERENTIAL EQUATIONS with Computing and Modeling

Some computer algebra systems permit the use of unit step functions for the efficient derivation of Fourier series of "piecewise-defined" functions. Let the "unit function" $unit(t,a,b)$ have the value 1 on the interval $a \le t < b$ and the value 0 otherwise. Then we can define a given piecewise smooth function $f(t)$ as a "linear combination" of different unit functions corresponding to the separate intervals on which the function is smooth, with the unit function for each interval multiplied by the formula defining $f(t)$ on that interval. For example, consider the even period 2π function whose graph is shown below. This "trapezoidal wave function" is defined for $0 < t < \pi$ by

$$f(t) \;=\; \left(\tfrac{\pi}{3}\right) unit\left(t, 0, \tfrac{\pi}{6}\right) + \left(\tfrac{\pi}{2} - t\right) unit\left(t, \tfrac{\pi}{6}, \tfrac{5\pi}{6}\right) + \left(-\tfrac{\pi}{3}\right) unit\left(t, \tfrac{5\pi}{6}, \pi\right). \qquad (1)$$

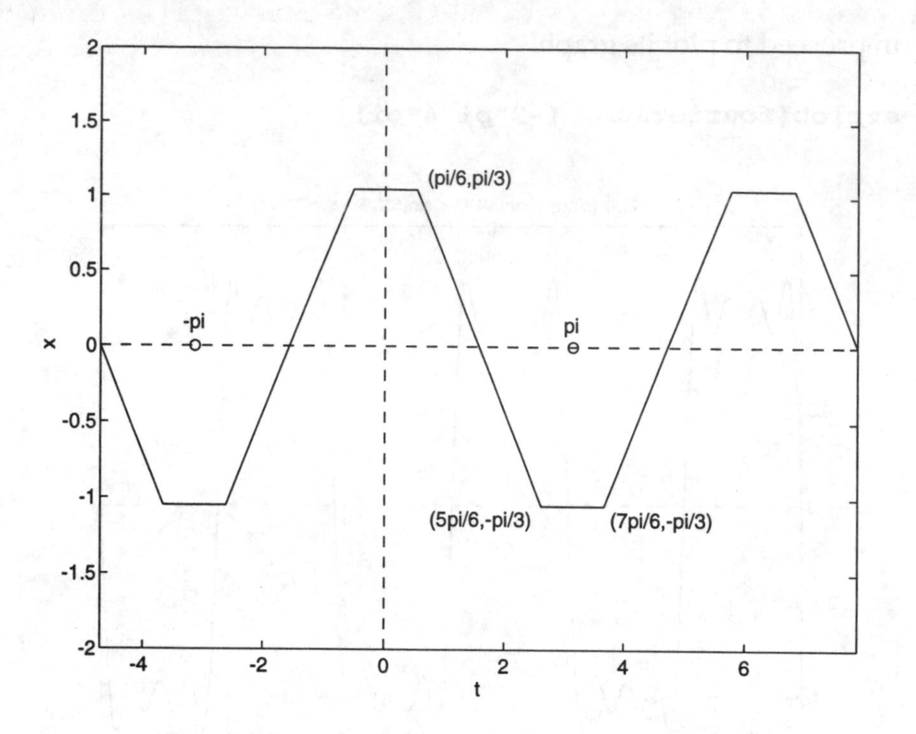

We can then substitute (1) in the Fourier cosine coefficient formula

$$a_n = \frac{2}{\pi}\int_0^{\pi} f(t)\cos nt\,dt \tag{2}$$

to discover the lovely Fourier series

$$f(t) = \frac{2\sqrt{3}}{\pi}\sum \frac{(\pm)\cos nt}{n^2} \tag{3}$$

with a $+--++--++$ ..... pattern of signs, the summation being taken over all odd positive integers n that are *not* multiples of 3.

In the paragraphs that follow we illustrate this approach using *Maple*, *Mathematica*, and MATLAB. You can then apply this method to find the Fourier series of the following period 2π functions.

1. The even square wave function whose graph is shown in Figure 9.3.8 in the text.

2. The even and odd triangular wave functions whose graphs are shown in Figures 9.2.4 and 9.3.9.

3. The odd trapezoidal wave function whose graph in shown in Figure 9.2.5.

Then find similarly the Fourier series of some piecewise smooth functions of your own choice, perhaps ones that have periods other than 2π and are neither even nor odd.

Using *Maple*

The unit step function (with values 0 for $t < 0$ and 1 for $t > 0$) is available in *Maple* as the "Heaviside function":

```
> Heaviside(-2);  Heaviside(3);
```

$$0$$
$$1$$

The "unit function" on the interval $[a, b]$ is then defined by

```
> unit := (t,a,b) -> Heaviside(t-a) - Heaviside(t-b):
```

You can

```
> plot( unit(t,1,2), t = 0..3 );
```

to verify that this unit function looks as intended. The trapezoidal wave function in (1) is then defined for $0 \le t \le \pi$ by

```
> f := t ->   (Pi/3)*unit(t,     0,    Pi/6)+
          (Pi/2 - t)*unit(t, Pi/6,   5*Pi/6) +
             (-Pi/3)*unit(t, 5*Pi/6,    Pi):
```

and the graph resulting from the command

```
> plot( f(t), t=0..Pi );
```

agrees with the previous figure. However, the integration function in *Maple* does not fully respect combinations of Heaviside functions — for instance, the command `int(f(t), t = 0..Pi)` fails to give the simplied result 0 that we expect (why?) — so we calculate the Fourier coefficients in (2) by piecewise integration.

```
> L := Pi:
> a := n ->
        (2/L)*( int(   (Pi/3)*cos(n*t), t=0..Pi/6) +
                int((Pi/2-t)*cos(n*t), t=Pi/6..5*Pi/6) +
                int( -(Pi/3)*cos(n*t), t=5*Pi/6..Pi) ):
```

The a typical partial sum of the Fourier cosine series of $f(t)$ is given by

```
> fourierSum :=
  a(0)/2 + sum( a(n)*cos(n*Pi*t/L), n=1..25 );
```

$$fourierSum := 2\frac{\sqrt{3}\cos(t)}{\pi} - \frac{2}{25}\frac{\sqrt{3}\cos(5\,t)}{\pi} - \frac{2}{49}\frac{\sqrt{3}\cos(7\,t)}{\pi} + \frac{2}{121}\frac{\sqrt{3}\cos(11\,t)}{\pi} + \frac{2}{169}\frac{\sqrt{3}\cos(13\,t)}{\pi} - \frac{2}{289}\frac{\sqrt{3}\cos(17\,t)}{\pi} - \frac{2}{361}\frac{\sqrt{3}\cos(19\,t)}{\pi} + \frac{2}{529}\frac{\sqrt{3}\cos(23\,t)}{\pi} + \frac{2}{625}\frac{\sqrt{3}\cos(25\,t)}{\pi}$$

in agreement with (3). You can enter the command

```
> plot(fourierSum, t=-5..8);
```

to verify that this Fourier series is consistent with the trapezoidal wave figure shown previously.

Using *Mathematica*

The unit step function (with values 0 for $t < 0$ and 1 for $t > 0$) is available for *Mathematica* calculations if we load the delta functions package:

```
<<Calculus`DiracDelta`
```

The "unit function" on the interval $[a, b]$ is then defined by

```
unit[t_,a_,b_] := UnitStep[t-a] - UnitStep[t-b]
```

You can

```
Plot[unit[t,1,2], {t,0,3} ]
```

to verify that this unit function looks as intended. The trapezoidal wave function in (1) is then defined $0 \le t \le \pi$ by

```
f[t_] :=      (Pi/3) unit[t, 0,        Pi/6] +
         (Pi/2 - t) unit[t, Pi/6, 5*Pi/6] +
            (-Pi/3) unit[t, 5*Pi/6,     Pi]
```

It is clear that $a_0 = 0$ (why?), as we verify with the computation

```
L  = Pi;
a0 = (2/L)*Integrate[ f[t], {t, 0, L} ]
```

```
0
```

The coefficient a_n is given for $n > 0$ by

```
a[n_] := (2/L)*Integrate[ f[t] Cos[n*Pi*t/L], {t,0,L} ]
```

and finally a typical partial sum of the Fourier cosine series of $f(t)$ is given by

```
fourierSum = a0/2 + Sum[ a[n]*Cos[n*Pi*t/L], {n,1,25} ]
```

```
 2 Sqrt[3] Cos[t]    2 Sqrt[3] Cos[5 t]    2 Sqrt[3] Cos[7 t]
 ---------------- - ------------------ - ------------------ +
        Pi                25 Pi                 49 Pi

  2 Sqrt[3] Cos[11 t]   2 Sqrt[3] Cos[13 t]
  ------------------- + ------------------- -
        121 Pi                169 Pi

  2 Sqrt[3] Cos[17 t]   2 Sqrt[3] Cos[19 t]
  ------------------- - ------------------- +
        289 Pi                361 Pi

  2 Sqrt[3] Cos[23 t]   2 Sqrt[3] Cos[25 t]
  ------------------- + -------------------
        529 Pi                625 Pi
```

in agreement with (3). You can enter the command

```
Plot[ Evaluate[fourierSum], {t,-5,8} ]
```

to verify that this Fourier series is consistent with the trapezoidal wave figure shown previously.

Using MATLAB

The "unit function" on the interval $[a, b]$ can be defined by in MATLAB by

```
function  y = unit(t,a,b)

y = (sign(t - a) - sign(t - b))/2;
```

You can enter the command

```
» t = 0 : pi/300 : pi;
» plot( t, unit(t,1,2) )
```

to verify that this unit function looks as intended. The trapezoidal wave function in (1) is then defined $0 \leq t \leq \pi$ by

```
function y = f(t)

y =      (pi/3)*unit(t, 0,pi/6) + ...
     (pi/2-t).*unit(t, pi/6,5*pi/6) + ...
       (-pi/3)*unit(t, 5*pi/6,pi);
```

and the graph resulting from the commands

```
» t = 0 : pi/300 : pi;
» plot(t, f(t))
```

agrees with the previous figure. However, the *Maple*-based symbolic integration function available in MATLAB does not respect combinations of unit functions, so we calculate the Fourier coefficients in (2) by piecewise integration. First we verify that the constant term in the Fourier cosine series of $f(t)$ is 0.

```
» I1 = int('pi/3', 0, pi/6);
» I2 = int('pi/2 - t', pi/6, 5*pi/6);
» I3 = int('-pi/3', 5*pi/6, pi);
» a0 = symop(I1,'+',I2,'+',I3)

a0 =
0
```

Then we use a loop to add successive terms in order to calculate a partial sum of the series.

```
» fourierSum = a0;

»for k = 1 : 25
     cosine = subs('cos(n*t)',k,'n');
     I1 = int( symmul('pi/3',cosine), 0,pi/6);
     I2 = int( symmul('pi/2-t',cosine), pi/6, 5*pi/6);
     I3 = int( symmul('-pi/3',cosine), 5*pi/6, pi);
     an = symop('2/pi','*','(',I1,'+',I2,'+',I3,')');
     term = symmul(an,cosine);
     fourierSum = symadd(fourierSum, term);
     end
```

The command

```
» pretty(fourierSum)
```

then displays the result:

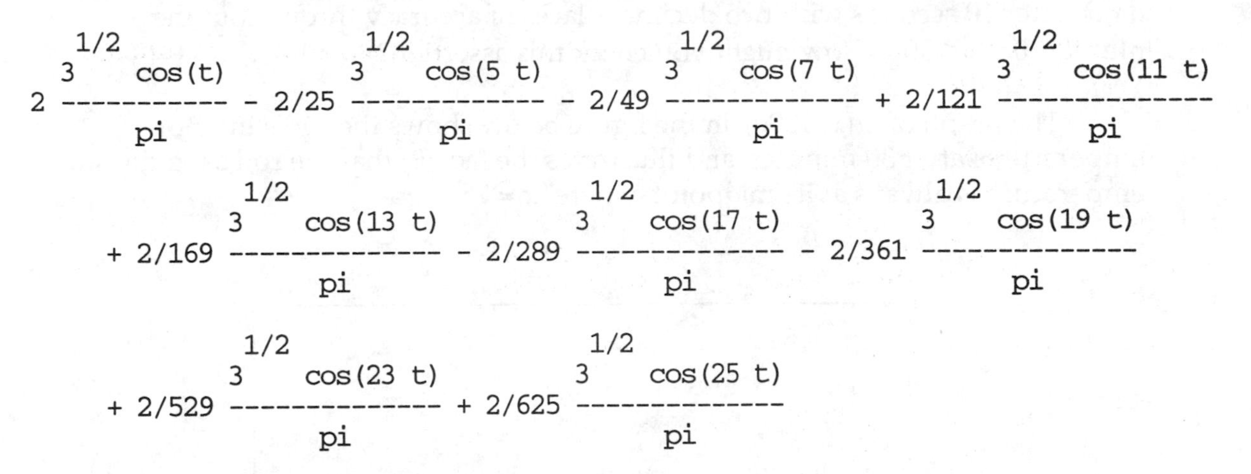

```
   1/2                    1/2                  1/2                      1/2
  3    cos(t)            3    cos(5 t)         3    cos(7 t)             3    cos(11 t)
2 ----------- - 2/25 ------------- - 2/49 ------------- + 2/121 --------------
      pi                    pi                    pi                       pi

             1/2                      1/2                     1/2
            3    cos(13 t)            3    cos(17 t)          3    cos(19 t)
    + 2/169 -------------- - 2/289 -------------- - 2/361 --------------
                  pi                     pi                      pi

             1/2                      1/2
            3    cos(23 t)            3    cos(25 t)
    + 2/529 -------------- + 2/625 --------------
                  pi                     pi
```

in agreement with (3). You can enter the command

```
» ezplot(fourierSum, [-3 8])
```

to verify that this Fourier series is consistent with the trapezoidal wave figure shown previously.

Project 38
Numerical Investigation of Heat Flow Problems

Reference: Section 9.5 of Edwards & Penney *DIFFERENTIAL EQUATIONS with Computing and Modeling*

In the paragraphs below we illustrate the use of *Maple, Mathematica*, and MATLAB to investigate numerically the temperature function

$$u(x,t) \;=\; \frac{4T_0}{\pi}\sum_{n\,odd}\frac{1}{n}\exp\left(-\frac{n^2\pi^2kt}{L^2}\right)\sin\frac{n\pi x}{L} \qquad (1)$$

of the heated rod of Example 2 in Section 9.5 of the text. This rod of length $L =$ 50 cm and thermal diffusivity $k = 0.15$ (for iron) has constant initial temperature $u(x, 0) = T_0 = 100°$ and zero endpoint temperatures $u(0, t) = u(\mathrm{L}, t)$ $= 0$ for $t > 0$.

Once we have defined a function that calculates partial sums of the series in (1), we can graph $u(x, t)$ in order to investigate the rate at which the initially heated rod cools. As a practical matter, $m = 25$ terms suffice to give the value

$u(x,t)$ after 10 seconds with two decimal places of accuracy throughout the interval $0 \leq x \leq 50$. (How might you check this assertion?)

The graph of $u(x, 1800)$ in the figure below shows the rod's interior temperatures after 30 minutes, and illustrates the fact (?) that the rod's maximum temperature is always at its midpoint, where $x = 25$.

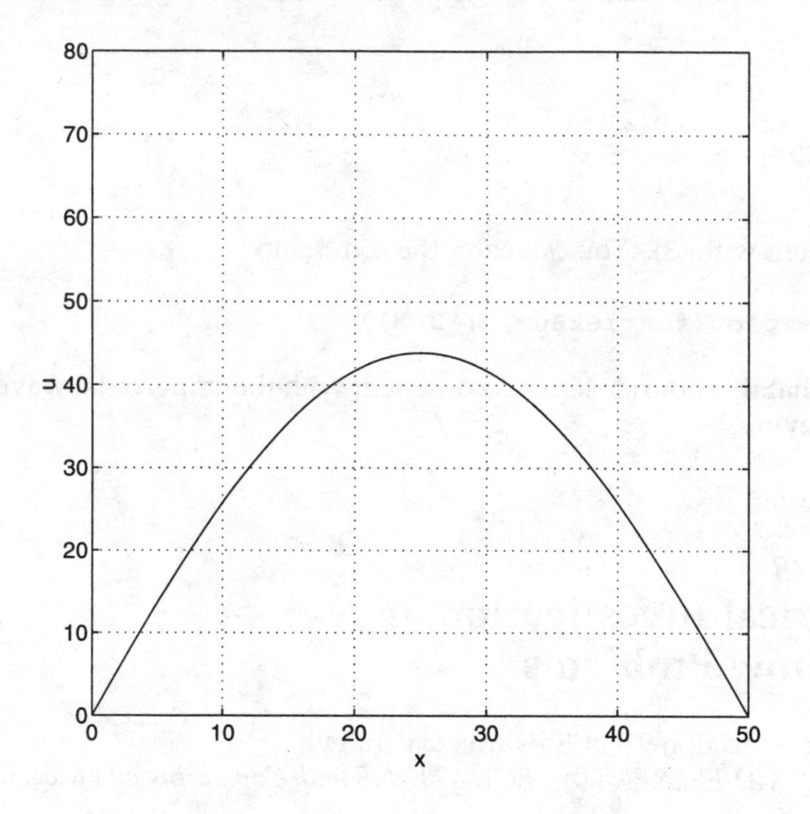

The next graph — of $u(25, t)$ for an initial two-hour period — indicates that the midpoint temperature takes something more than 1500 seconds (25 minutes) to fall to 50°. An appropriate magnification of this graph near its intersection point with the horizontal line $u = 50$ indicates that this actually takes about 1578 seconds (26 min 18 sec).

For your very own rod with constant initial temperature $u(x, 0) = T_0 =$ 100° to investigate in this manner, let

$$L = 100 + 10p \quad \text{and} \quad k = 1 + (0.1)q$$

where p is the largest and q is the smallest nonzero digit of your student I.D. number.

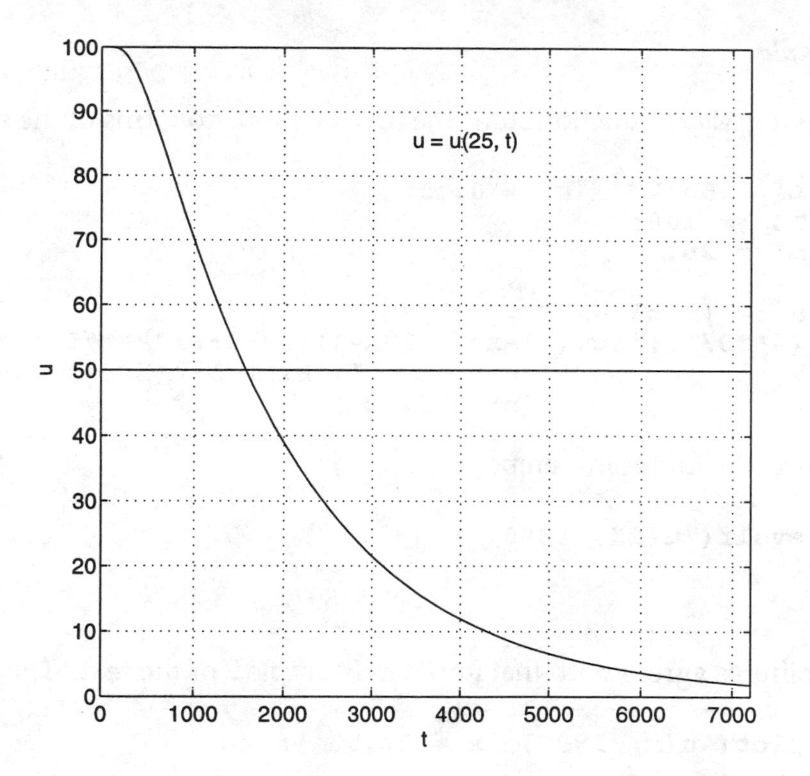

Investigation 1
Investigate generally the evolution of temperatures in the rod. Show graphs of u versus x for selected values of t. Then determine how long it will be until the maximum temperature anywhere in the rod is 50 degrees You might begin graphically with an appropriate graph of u versus t.

Investigation 2
If the end $x = 0$ of the rod is held at temperature 0, while the end $x = L$ is insulated, then Problem 24 in Section 9.5 says that its temperature function $u(x,t)$ is given by

$$u(x,t) = \sum_{n \text{ odd}}^{\infty} c_n \exp\left(-\frac{n^2\pi^2 kt}{4L^2}\right) \sin\left(\frac{n\pi x}{2L}\right) \tag{2}$$

where

$$c_n = \frac{2}{L}\int_0^L f(x) \sin\frac{n\pi x}{2L}\, dx \tag{3}$$

for $n = 1, 3, 5, \ldots$. Investigate generally the evolution of temperatures in the rod. Show graphs of u versus x for selected values of t. Then determine how long it will be until the maximum temperature anywhere in the rod is 50 degrees.

Using *Maple*

The following *Maple* function sums the first m nonzero terms of the series in (1).

```
> L := 50:        k := 0.15:
  T0 := 100:
  m := 25:

> u := (x,t) ->
  (4*T0/Pi)*sum( 'exp(-(2*n-1)^2*Pi*Pi*k*t/L^2)*
                  sin((2*n-1)*Pi*x/L)/(2*n-1)',
                 'n' = 1..m );
```

For instance, the midpoint temperature

```
> evalf( u(25, 1800) );
```

43.848977

after 30 minutes agrees with that found in Example 2 of the text. The commands

```
> plot( u(x, 1800), x = 0..50 );

> plot( u(25, t), t = 0..7200 );
```

produce graphs like those shown previously.

Using *Mathematica*

The following *Mathematica* function sums the first m nonzero terms of the series in (1).

```
L = 50;        k = 0.15;
T0 = 100;
m = 25;

u[x_,t_] :=
(4*T0/Pi)*Sum[ Exp[-n*n*Pi*Pi*k*t/L^2]*Sin[n*Pi*x/L]/n,
                 {n, 1, 2*m -1, 2} ] // N
```

For instance, the midpoint temperature

```
u[ 25, 1800 ]

43.849
```

after 30 minutes agrees with that found in Example 2 of the text. The commands

```
Plot[ u[x, 1800], {x, 0, 50} ]

Plot[ u[25, t], {t, 0, 7200} ]
```

produce graphs like those shown previously.

Using MATLAB

The following MATLAB function sums the first m nonzero terms of the series in (1).

```
function   u = u(x,t)

L = 50;                   % length of rod
k = 0.15;                 % thermal diffusivity of rod
T0 = 100;                 % constant initial temp of rod
m = 25;                   % no of nonzero terms to use

n = 2*(1:m) - 1;          % row vector of m odd indices

coeffs = 4*T0 ./(pi*n);          % row vector of m coeffs
coeffs = diag(coeffs);           % m x m  diagonal matrix

exps = exp(-(n'.*n')*pi^2*k*t/L^2); %  m x p  matrix
sines = sin(pi*n'*x/L);             %  m x q  matrix
u = exps'*coeffs*sines;             %  p x q  matrix
```

(Note how we use matrix multiplication for fast summation.) For instance, the midpoint temperature

```
» u(25,1800)
ans =
   43.8490
```

after 30 minutes agrees with that found in Example 2 of the text. The commands

```
» x = 0 : 0.1 : 50;
» plot(x, u(x,1800))

» t = 0 : 20 : 7200;
» plot(t, u(25,t))
```

produce graphs like those shown previously.

Project 39

Computer Implementation of d'Alembert's Method

Reference: Section 9.6 of Edwards & Penney
DIFFERENTIAL EQUATIONS with Computing and Modeling

The d'Alembert solution

$$y(x,t) \quad = \quad \tfrac{1}{2}[F(x+at)+F(x-at)] \tag{1}$$

of the vibrating string problem (with fixed endpoints and zero initial velocity on the interval $[0, L]$) is readily implemented in a computer system such as *Maple*, *Mathematica*, or MATLAB. Recall that $F(x)$ in (1) denotes the *odd* period 2L extension of the string's initial position function $f(x)$. A plot of $y = y(x,t)$ for $0 \le x \le L$ with t fixed shows a snapshot of the string's position at time t.

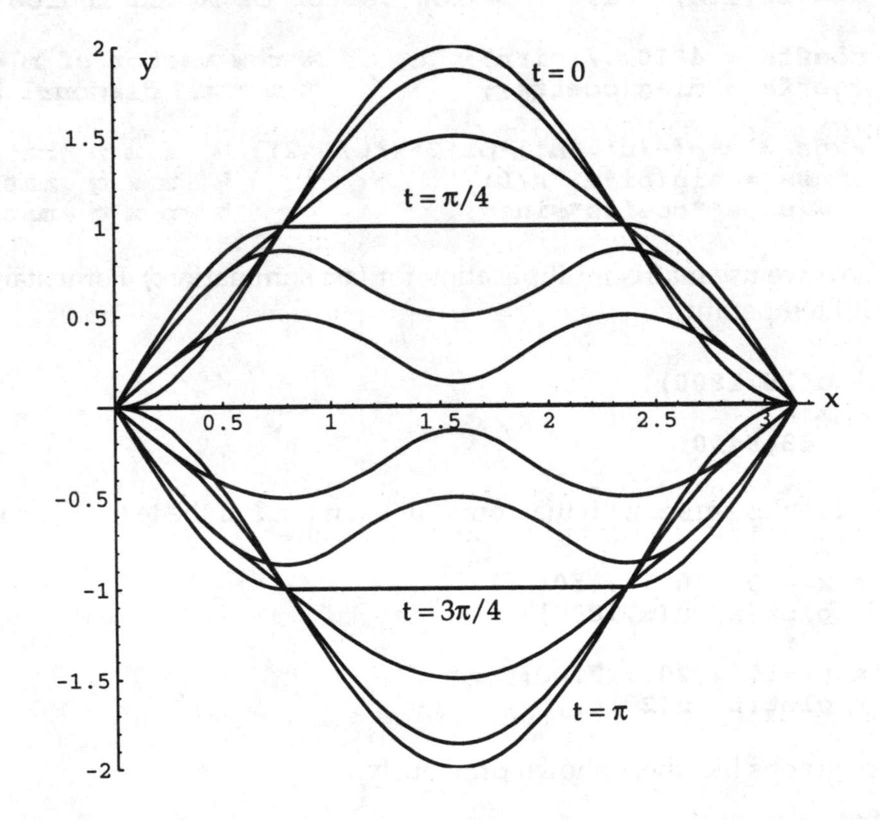

In the paragraphs that follow we illustrate the use of *Maple*, *Mathematica*, and MATLAB to plot such snapshots for the vibrating string with initial position function

$$f(x) = 2\sin^2 x, \qquad 0 \le x \le \pi. \tag{2}$$

We can illustrate the motion of this vibrating string by plotting a sequence of snapshots , either separately (as in Fig. 9.6.4 of the text) or on a single figure — as in the preceding figure. The apparent "flat spots" in the $t = \pi/4$ and $t = 3\pi/4$ snapshots are discussed in Problem 23 of Section 9.6 in the text.

You can test your implementation of d'Alembert's method by attempting to generate Figures 9.6.6 (for a string with triangular initial position) and 9.6.7 (for a string with trapezoidal initial position) in the text. The initial position "bump function"

$$f(x) = \sin^{200} x, \qquad 0 \le x \le \pi. \tag{3}$$

generates travelling waves traveling (initially) in opposite directions, as indicated in Fig. 9.6.3 in the text. The initial position function defined by

$$f(x) = \begin{cases} \sin^{200}(x+1.5) & \text{for } 0 < x < \pi/2, \\ 0 & \text{for } \pi/2 < x < \pi. \end{cases} \tag{4}$$

generates a single wave that starts at $x = 0$ and (initially) travels to the right. (Think of a jump rope tied to a tree, whose free end is initially "snapped".)

After exploring some of the possibilities indicated above, try some initial position functions of you own choice. Any continuous function f such that $f(0) = f(L) = 0$ is fair game. The more exotic the resulting vibration of the string, the better.

Using *Maple*

To plot the snapshots plotted simultaneously in the preceding figure, we started with the string's initial position function $f(x)$ defined by

```
> f := x -> 2*sin(x)^2:
```

To define the odd period 2π extension $F(x)$ of $f(x)$, we need the function $s(x)$ that shifts the point x by a multiple of π into the interval $[0, \pi]$.

```
> s := proc(x)
       local k;
       k := floor(evalf(x/Pi)):
       if type(k, even)
            then evalf(x - k*Pi):
            else evalf(x - k*Pi - Pi) fi
       end:
```

Then the desired odd extension is defined by

```
> F := proc(x)
        if s(x) > 0 then    f(s(x))
                     else -f(s(-x)) fi
        end:
```

Finally, the d'Alembert solution in (1) is

```
> G := (x,t) -> ( F(x + t) + F(x - t) )/2:
```

We now expect the the command

```
> t := Pi/4:  plot( G(x,t), x = 0..Pi );
```

to plot the $t = \pi/4$ snapshot exhibiting the apparent flat spot previously mentioned. Instead, it produces the message

```
Error, (in F) cannot evaluate boolean
```

The usual trick to prevent premature evaluation is

```
> plot( 'G(x,t)', x = 0..Pi );
```

but this produces an obviously erroneous figure — try it and see.

The solution is to plot individual points on the desired graph and connect these points with line segments. If we want to plot the graphs for

$$t = 0, \pi/12, \pi/6, \pi/4, \pi/4, 5\pi/12, \cdots, \pi,$$

a sequence of 201 points on the nth of these graphs is defined by

```
> plotpoints := n ->
  [seq( [i*Pi/200, G(i*Pi/200,n*Pi/12)], i=0..200 )]:
```

The nth of our plots is then defined by

```
> fig := n -> plot( plotpoints(n),
                    x = 0..Pi, y = -2..2,
                    style = line ):
```

The preceding figure, exhibiting simultaneously the successive positions of the string in a single composite figure, is finally generated by the command

```
> with(plots):
  display( [seq( fig(n), n = 0..12) ] );
```

The initial position function

```
> f := proc(x)
       if x < Pi/2 then x else Pi - x fi
       end:
```

corresponding to the triangular wave function of Project 36 generates in this way the composite picture shown in Fig 9.6.6 of the text. Similarly, the trapezoidal wave function

```
> f := proc(x)
       if x <= Pi/3 then x
          elif  x > Pi/3 and x < 2*Pi/3 then Pi/3
          else Pi - x  fi
       end:
```

of Project 36 produces the picture shown in Fig. 9.6.7.

Using *Mathematica*

To plot the snapshots plotted simultaneously in the preceding figure, we started with the string's initial position function $f(x)$ defined by

```
f[x_] := 2 Sin[x]^2
```

To define the odd period 2π extension $F(x)$ of $f(x)$, we need the function $s(x)$ that shifts the point x by a multiple of π into the interval $[0, \pi]$.

```
s[x_] := Block[{k}, k = Floor[ N[x/Pi] ];
                 If[EvenQ[k], (* k is even *)
                 (* then *)    N[x - k*Pi],
                 (* else *)    N[x - k*Pi - Pi]] ]
```

Then the desired odd extension is defined by

```
F[x_] := If[ s[x] > 0, (* then *)   f[ s[x]],
                       (* else *)  -f[-s[x]] ]
```

Finally, the d'Alembert solution in (1) is

```
G[x_,t_] := ( F[x + t] + F[x - t] )/2
```

A snapshot of the position of the string at time t is plotted by

```
stringAt[ t_ ] := Plot[ G[x,t], {x,0,Pi},
                        PlotRange -> {-2,2} ];
```

For example, the command

```
stringAt[ Pi/4 ]
```

plots the $t = \pi/4$ snapshot exhibiting the apparent flat spot previously mentioned. We can plot a whole sequence of snapshots at once:

```
snapshots = Table[ stringAt[t], {t, 0, Pi, Pi/12} ]
```

These snapshots can be animated to show the vibrating string in motion, or we can exhibit simultaneously the successive positions of the string in a single composite figure with the command

```
Show[ snapshots ]
```

The initial position function

```
f[x_] := If[ x < Pi/2, (* then *) x,
                       (* else *) Pi - x ] // N
```

corresponding to the triangular wave function of Project 36 generates in this way the composite picture shown in Fig 9.6.6 of the text. Similarly, the trapezoidal wave function

```
f[x_] := Which[      0 <= x < Pi/3,       x,
                  Pi/3 <= x < 2*Pi/3,  Pi/3,
                2*Pi/3 <= x <= Pi,     Pi - x ] // N
```

of Project 36 produces the picture shown in Fig. 9.6.7.

Using MATLAB

To plot the snapshots plotted simultaneously in the preceding figure, we started with the string's initial position function $f(x)$ defined by

```
function  y = f(x)
y = 2*sin(x).^2;
```

To define the odd period 2π extension $F(x)$ of $f(x)$, we need first to shift the point x by a multiple of 2π to a point s in the interval $[-\pi, \pi]$. We then define $F(x)$ to be $f(s)$ if $s > 0$, $-f(-s)$ if $x < 0$. This is accomplished by the function

```
function y = foddext(x)

%  Odd period 2Pi extension of the function f

k = floor(x/pi);
```

```
q = ( 2*floor(k/2) ~= k );    % q = 0 if k even
s = x - (k+q)*pi;             % q = 1 if k odd
m = sign(s);

%  if s>0 then y = f(s) else y = -f(-s)
y = m.*f(m.*s);
```

Finally, the d'Alembert solution in (1) is defined by

```
function  y = G(x,t)
y = ( foddext(x+t) + foddext(x-t) )/2;
```

Then the commands

```
» x = 0 : pi/200 : pi;
» plot(x, G(x,pi/4))
```

plot the $t = \pi/4$ snapshot exhibiting the apparent flat spot previously mentioned. The simple loop

```
» for n = 0 : 12
      plot( x, G(x, n*pi/12) )
      axis([0 pi -2 2]); hold on
      end
```

finally generates the preceding composite figure that exhibits simultaneously the successive positions of the vibrating string from $t = 0$ to $t = \pi$.

The initial position function

```
function   y = f3(x)
y = x.*(x < pi/2) + (pi-x).*(x >= pi/2);
```

corresponding to the triangular wave function of Project 36 generates in this way the composite picture shown in Fig 9.6.6 of the text. Similarly, the trapezoidal wave function

```
function  y = f(x)
y = x.*(x <= pi/3) ...
    + (pi/3)*(x > pi/3 & x < 2*pi/3)...
    + (pi - x).*(x >= 2*pi/3);
```

of Project 36 produces the picture shown in Fig. 9.6.7.

Chapter 10

Eigenvalues and Boundary Value Problems

Project 40
Heat Transfer Investigations

Reference: Section 10.2 of Edwards & Penney
DIFFERENTIAL EQUATIONS with Computing and Modeling

In the paragraphs that follow we illustrate the use of *Maple*, *Mathematica*, and MATLAB to investigate numerically the temperature function

$$u(x,t) \;=\; 2u_0hL\sum_{n=1}^{\infty}\frac{1-\cos\beta_n}{\beta_n(hL+\cos^2\beta_n)}\exp\left(-\frac{\beta_n{}^2kt}{L^2}\right)\sin\frac{\beta_n x}{L} \qquad (1)$$

of the heated slab of Example 1 in Section 10.2 of the text. We suppose that this slab has thickness $L = 50$ cm, uniform initial temperature $u_0 = 100°$, fixed temperature 0° at the slab's left-hand boundary $x = 0$, thermal diffusivity $k =$ 0.15 (for iron), and heat transfer coefficient $h = 0.1$ at the right-hand boundary $x = L$.

According to Eq. (8) in the text, the eigenvalues $\{\beta_n\}_1^\infty$ in (1) are the positive solutions of the equation $\tan x \;=\; -x\,/\,hL$. Figure 10.1.1 in the text exhibits these eigenvalues as the x-coordinates of the points of intersection of $y \;=\; x\,/\,hL$ and $y \;=\; -\tan x$. There we see that if n is large then β_n is slightly larger than $(2n-1)\pi\,/\,2$. This observation provides initial estimates with which we can proceed to approximate the eigenvalues numerically (using Newton's method, for instance). Once a sufficient number of the $\{\beta_n\}_1^\infty$ have been found, we can proceed to sum the series (1) numerically — as needed to plot the temperature u either as a function of x or as a function of t.

The first figure below illustrates the falling temperature of the slab at its right-hand boundary where heat transfer with a medium at temperature 0° takes place. This graph of $u \;=\; u(50,t)$ for the first hour indicates that the right-hand boundary temperature of the slab falls to 25° in a bit less than 2000 seconds. We will see that this actually takes about $t_1 = 1951$ seconds. The graph $u \;=\; u(x,t_1)$

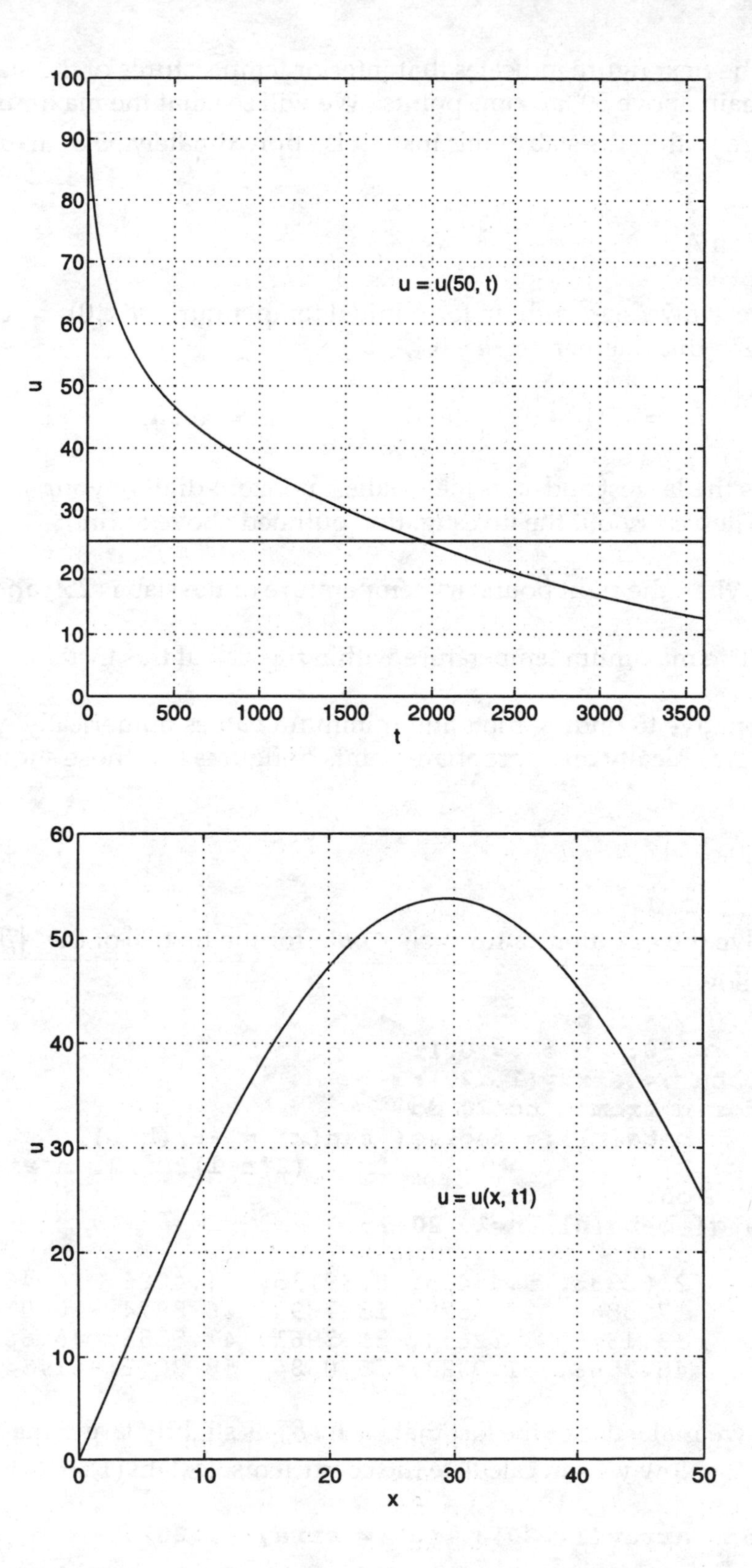

100
90
80
70
60
50
40
30
20
10
0
u
u = u(50, t)
0
500
1000
1500
2000
2500
3000
3500
t
60
50
40
30
20
10
0
u
u = u(x, t1)
0
10
20
30
40
50
x

shown in the next figure indicates that interior temperatures of the slab at time t_1 still remain above 50° at some points. We will see that the maximum temperature within the slab at this instant is approximately 53.8° at $x \approx 29.4$.

Investigation A

For your very own slab with uniform initial temperature $u(x,0) = 100°$ to investigate in this manner, let $h = 0.1$,

$$L = 5(10+p) \qquad \text{and} \qquad k = 0.1q,$$

where p is the largest and q is the smallest nonzero digit of your student I.D. number. Then carry out the investigation outlined above to find:

- When the right boundary temperature of the slab is 25°; and
- The maximum temperature within the slab at this instant.

As an alternative to finding roots and minimum values numerically, you can "zoom in" graphically on appropriate points of figures like those shown above.

Using *Maple*

We can solve the equation $\tan x = -x/hL$ for the first 20 of the $\{\beta_n\}_1^\infty$ values in (1) as follows:

```
> L := 50:     h := 0.1:
> beta := array(1..20):
> for n from 1 to 20 do
      beta[n] := fsolve( tan(x) = -x/(h*L),
                              x = (2*n-1)*Pi/2..n*Pi ):
      od:
> seq( beta[n], n=1..20 );

     {2.65366, 5.45435, 8.39135, 11.4086, 14.4699,
      17.5562, 20.6578, 23.7693, 26.8874, 30.0102,
      33.1365, 36.2653, 39.3961, 42.5285, 45.6622,
      48.7968, 51.9323, 55.0684, 58.2052, 61.3424}
```

Note how we make use of the fact that each β_n is slightly larger than $(2n-1)\pi/2$. Now we can calculate the coefficients $\{c_n\}$ in (1):

```
> b : array(1..20):    c := array(1..20):
```

```
> for n from 1 to 20 do
      b[n]  := beta[n]:
      c[n]  := (1-cos(b[n]))/(b[n]*(h*L+cos(b[n])^2)):
      od:
```

The following function then sums the first 20 nonzero terms of the series in Eq. (1).

```
> u0 := 100:       k := 0.15:      n := 'n':
> u := (x,t) -> 2*u0*h*L*
                  sum( c[n]*exp(-b[n]*b[n]*k*t/L^2)*
                            sin(b[n]*x/L), n = 1..10 );
```

As a practical matter, this suffices to calculate the value $u(x,t)$ after 10 seconds with two-place accuracy throughout the interval $0 \le x \le 50$. (How might you verify this assertion?)

The command

```
> plot( {u(50,t), 25}, t=0..3600 );
```

yields the preceding graph of $u(50,t)$ for $0 \le t \le 3600$. The solution

```
> t1 := fsolve( u(50,t) = 25, t = 1900..2000 );

                          t1 := 1951.24
```

then shows that it takes just over $t_1 = 1951$ seconds for the right-hand boundary temperature of the slab to fall to 25°. The command

```
> plot( u(x,t1), x = 0..50 );
```

graphs $u = u(x,t_1)$ for $0 \le x \le 50$, and— observing the maximum near $x = 30$ — we finally maximize the temperature $u(x,t1)$:

```
> maximize( u(x,t1), x, 29..31 );

                             53.7942
```

Thus the maximum temperature within the slab at time $t = t_1$ is just under 54°.

Using *Mathematica*

We can solve the equation $\tan x = -x / hL$ for the first 20 of the $\{\beta_n\}_1^\infty$ values in (1) as follows:

```
L = 50;        h = 0.1;
solutions =
Table[ FindRoot[ Tan[x] == -x/(h*L),
                 {x, (2*n - 1)*Pi/2 + 0.1} ],
       {n, 1, 20} ];
beta = x /. solutions

{2.65366, 5.45435, 8.39135, 11.4086, 14.4699,
 17.5562, 20.6578, 23.7693, 26.8874, 30.0102,
 33.1365, 36.2653, 39.3961, 42.5285, 45.6622,
 48.7968, 51.9323, 55.0684, 58.2052, 61.3424}
```

Note how we make use of the fact that each β_n is slightly larger than $(2n-1)\pi/2$. The following function then sums the first 20 nonzero terms of the series in Eq. (1).

```
b = beta;     u0 = 100;     k = 0.15;
c = (1 - Cos[b])/(b*(h*L + Cos[b]^2));      (* coeffs *)
u[x_,t_] :=
2*u0*h*L*Apply[ Plus,
                c*Exp[-b*b*k*t/L^2]*Sin[b*x/L] ] // N
```

As a practical matter, this suffices to calculate the value $u(x,t)$ after 10 seconds with two-place accuracy throughout the interval $0 \le x \le 50$. (How might you verify this assertion?)

The command

```
Plot[ {u[50,t],25}, {t,0,3600} ]
```

yields the preceding graph of $u(50,t)$ for $0 \le t \le 3600$. The solution

```
t1 = t /. FindRoot[ u[50,t] == 25, {t, 2000} ]

1951.24
```

then shows that it takes just over $t_1 = 1951$ seconds for the right-hand boundary temperature of the slab to fall to 25°. The command

```
Plot[ u[x,t1], {x,0,50} ]
```

graphs $u = u(x,t_1)$ for $0 \le x \le 50$, and finally we can find the maximum value of the temperature $u(x,t1)$ by finding the minimum value of its negative $-u(x,t)$:

```
FindMinimum[ -u[x,t1], {x,30} ]

{-53.804, {x -> 29.3586}}
```

Thus the maximum temperature within the slab at time $t = t_1$ is just under 54°.

Using MATLAB

First we define the eigenvalue equation $\tan x = -x / hL$ we need to solve in the form

```
function  y = f(x)
L = 50;   h = 0.1;
y = h*L*sin(x) + x.*cos(x);
```

Then we proceed to find the first 20 of the $\{\beta_n\}_1^\infty$ values in (1) as follows:

```
» for n = 1 : 20
      b(n) = fzero('f', (2*n-1)*pi/2 + 0.1);
      end

» reshape(b,5,4)'
ans =
     2.6537      5.4544      8.3913     11.4086     14.4699
    17.5562     20.6578     23.7693     26.8874     30.0102
    33.1365     36.2653     39.3961     42.5285     45.6622
    48.7968     51.9323     55.0684     58.2052     61.3424
```

Note how we make use of the fact that each β_n is slightly larger than $(2n-1)\pi / 2$. Now we can calculate the coefficients $\{c_n\}$ in (1):

```
» c = (1 - cos(b))./(b.*(h*L + cos(b).^2));
```

Given the slab's initial temperature and thermal diffusivity

```
» u0 = 100;    k = 0.15;
```

we can proceed to sum the series in (1) to find the temperature at a given point x and time t. For instance, the calculation

```
» x = 50;  t = 3600;
» u = 2*u0*h*L*sum( c.*exp(-b.*b*k*t/L^2).*sin(b*x/L) )
u =
    12.5627
```

reveals that after 1 hour the slab's right-hand boundary temperature has fallen to about 12.56°. As a practical matter, this suffices to calculate the value $u(x,t)$ with two-place accuracy if $0 \le x \le 50$ and $t \ge 10$. (How might you verify this assertion?)

However, for plotting purposes it is better to assemble the commands above into a function `u(x,t)` that accepts as inputs a q-vector `x` and a p-vector `t` and returns a $p \times q$ matrix `u` of corresponding temperatures:

```
function    u = u(x,t)

%  x = row vector of  q  points of rod
%  t = row vector of  p  times
%  u = p x q  matrix of temperatures

L = 50;                         % length of rod
h = 0.1;                        % heat transfer coeff
u0 = 100;                       % initial temp of slab
k = 0.15;                       % thermal diffusivity
N = 20;                         % no of terms to use

for  n = 1 : N                  % solve for beta values
     b(n) = fzero('f', (2*n-1)*pi/2 + 0.1);
     end

c = 2*u0*h*L*(1 - cos(b))./(b.*(h*L + cos(b).^2));
coeffs = diag(c);                   %  N x N  diagonal
                                    % matrix of coeffs
exps = exp(-(b'.*b')*k*t/L^2);      %  N x p  matrix
sines = sin(b'*x/L);                %  N x q  matrix
u = exps'*coeffs*sines;             %  p x q  matrix
```

The final line of `u.m` uses matrix multiplication to sum the terms of the series. The commands

```
» t = 0 : 10 : 3600;
» plot( t, u(50,t) ), hold on
» plot( [0 3600], [25 25] )
```

then yield the preceding graph of $u(50,t)$ for $0 \le t \le 3600$. If we define the temperature function

```
function y = u50(t)
y = u(50,t) - 25;
```

giving the slab's right-hand boundary temperature at time t, then the solution

```
» t1 = fzero('u50', 2000)
t1 =
   1.9512e+03
```

then shows that it takes just over $t_1 = 1951$ seconds for the right-hand boundary temperature of the slab to fall to 25°. The command

```
» x = 0 : 0.2 : 50;  plot( x, u(x,t1) )
```

graphs $u = u(x,t_1)$ for $0 \leq x \leq 50$, and finally we can find the maximum value of the temperature $u(x,t1)$ by finding the minimum value of its negative $-u(x,t)$:

```
function  y = ut1(x)
t1 = 1951;
y = - u(x,t1);

»x1 = fmin('ut1', 29,31);
»u(x1,t1)
ans =
    53.8040
```

Thus the maximum temperature within the slab at time $t = t_1$ is just under 54°.

Investigation B

Here you are to investigate the cooling of a spherical ball of radius a that initially has constant temperature $u_0 = 100^\circ$ throughout. At time $t = 0$ it is immersed in a surrounding medium at temperature 0. At time t the temperature u in the ball then depends only on the distance r from its center, and $u(r,t)$ satisfies the boundary value problem

$$(r\,u)_t = k\,(r\,u)_{xx} \qquad (2)$$

$$H\,u(a,t) + K\,u_r(a,t) = 0 \qquad \text{(heat transfer at boundary)}$$

$$u(r,0) = u_0 \qquad \text{(initial temperature).}$$

The heat transfer condition includes the insulated case $H = 0$ which reduces to $u_r(a,t) = 0$, and the zero boundary temperature case $H = \infty$ which (for simple listing of cases) we take to mean that $u(a,t) = 0$.

To get started, substitute $v(r,t) = r\,u(r,t)$ in (2) to derive the transformed boundary value problem

$$v_t = k\,v_{rr} \qquad (0 < r < a) \qquad (3)$$

$$v(0,t) = 0 \qquad \text{(left endpoint)}$$

$$(H - K/a)\,v(a,t) + K\,v_r(a,t) = 0 \qquad \text{(right endpoint)}$$

$$v(r,0) = r\,u_0$$

involving the standard 1-dimensional heat equation in (3).

Show first that the separation of variables $v(r,t) = R(r)\,T(t)$ leads to an eigenvalue problem of the form

$$y''(x) + \lambda\, y(x) = 0 \qquad (0 < x < L)$$

$$y(0) = 0, \qquad h\, y(L) + y'(L) = 0$$

(in standard x, y, L notation). Solve this problem in the six cases

Case 0	$h = \infty$	(that is, $y(L) = 0$)
Case 1	$h > 0$	
Case 2	$h = 0$	(that is, $y'(L) = 0$)
Case 3	$-1 < hL < 0$	
Case 4	$hL = -1$	
Case 5	$hL < -1$	

In each case determine *all* the eigenvalues and eigenfunctions (including whether there are any zero or negative eigenvalues). In cases having non-standard eigenvalue equations, draw sketches showing the approximate locations of the eigenvalues, and approximate accurately the first the first several of them. In the remainder of the project you may need the first 40 or 50 eigenvalues for each case, but after a while you can get additional ones by simply adding π successively.

Now for the fun! With this mathematical preparation, you're ready to investigate the cooling of the physical ball. Suppose it's made of a copper alloy with

radius	$a = 10$ cm
density	$\delta = 10\ \text{gm/cm}^3$
specific heat	$c = 0.1\ \text{cal/gm}$
conductivity	$K = 1$ (in cgs units)

You are to analyze the following 5 cases:

Case 0	$H = \infty$	that is, $u(a,t) = 0$)
Case 1	$H = 0.5$	
Case 2	$H = 0.1$	
Case 3	$H = 0.05$	
Case 4	$H = 0$	(that is, $u_r(a,t) = 0$)

In each case, derive an infinite series solution of the form

$$u(r,t) = \sum c_n \exp\left(-\frac{\beta_n^2 kt}{a^2}\right)\frac{R_n(r)}{r}.$$

in terms of the eigenvalue-eigenfunction $\{\beta_n, R_n(t)\}$ pairs for the appropriate eigenvalue problem. Evaluate the appropriate integrals to express c_n in terms of $H, a,$ and β_n so you can calculate $u(r,t)$ numerically. Produce one or more *ru*-graphs illustrating the cooling within the ball. Finally, use *tu*-graphs to find how long it takes the center of the ball to cool to 50° (except in Case 4 where this makes no sense -- explain why).

Project 41
Vibrations of Elastic Bars

Reference: Section 10.3 of Edwards & Penney
DIFFERENTIAL EQUATIONS with Computing and Modeling

In this project you are to investigate further the vibrations of an elastic bar or beam of length L whose position function $y(x,t)$ satisfies the partial differential equation

$$\rho\frac{\partial^2 y}{\partial t^2} + EI\frac{\partial^4 y}{\partial x^4} = 0 \qquad (0 < x < L) \qquad (1)$$

and the initial conditions $y(x,0) = f(x), \;\; y_t(x,0) = 0.$

First separate the variables (as in Example 3 of Section 10.3 of the text) to derive the formal series solution

$$y(x,t) = \sum_{n=1}^{\infty} c_n X_n(x) \cos \frac{\beta_n^2 a^2 t}{L^2} \tag{2}$$

where $a^4 = EI/\rho$, the $\{c_n\}$ are the appropriate eigenfunction expansion coefficients of the initial position function $f(x)$, and the $\{\beta_n\}$ values and $\{X_n(x)\}$ eigenfunctions are determined by the end conditions imposed on the bar. In a particular case, one wants to find both the **frequency equation** whose positive roots are the $\{\beta_n\}$ and the explicit eigenfunctions $\{X_n(x)\}$. In this section we saw that the frequency equation for the *fixed-fixed* case (with $y(0) = y'(0) = y(L) = y'(L) = 0$) is

$$\cosh x \cos x = 1. \tag{3}$$

Case 1: Hinged at each end

The endpoint conditions are

$$y(0) = y''(0) = y(L) = y''(L) = 0. \tag{4}$$

According to Problem 8 in the text, the frequency equation is $\sin x = 0$, so

$$\beta_n = n\pi \quad \text{and} \quad X_n(x) = \sin n\pi x$$

for $n = 1, 2, 3, \ldots$. Suppose that the bar is made of steel (with density $\delta = 7.75$ g/cm^3 and Young's modulus $E = 2 \times 10^{12}$ dyn/cm^2), and is 19 inches long with square cross section of edge $w = 1$ in. (so its moment of inertia is $I = \frac{1}{12}w^4$). Determine its first few natural frequencies of vibration (in Hz). How does this bar sound when it vibrates?

Case 2: Free at each end

The endpoint conditions are

$$y''(0) = y'''(0) = y''(L) = y'''(L) = 0. \tag{5}$$

This case models, for example, a weightless bar suspended in space in an orbiting spacecraft. Now show that the frequency equation is $\cosh x \cos x = 1$ as in (3), although the eigenfunctions in this *free-free* case differ from those in the fixed-fixed case discussed in the text). From the figure showing the graphs $y = \operatorname{sech} x$ and $y = \cos x$ we see that $\beta_n \approx (2n+1)\pi/2$ for n large. Use the numerical methods of the Project 40 project to approximate the first several natural frequencies of vibration of the same physical bar considered in case 1. How does it sound now?

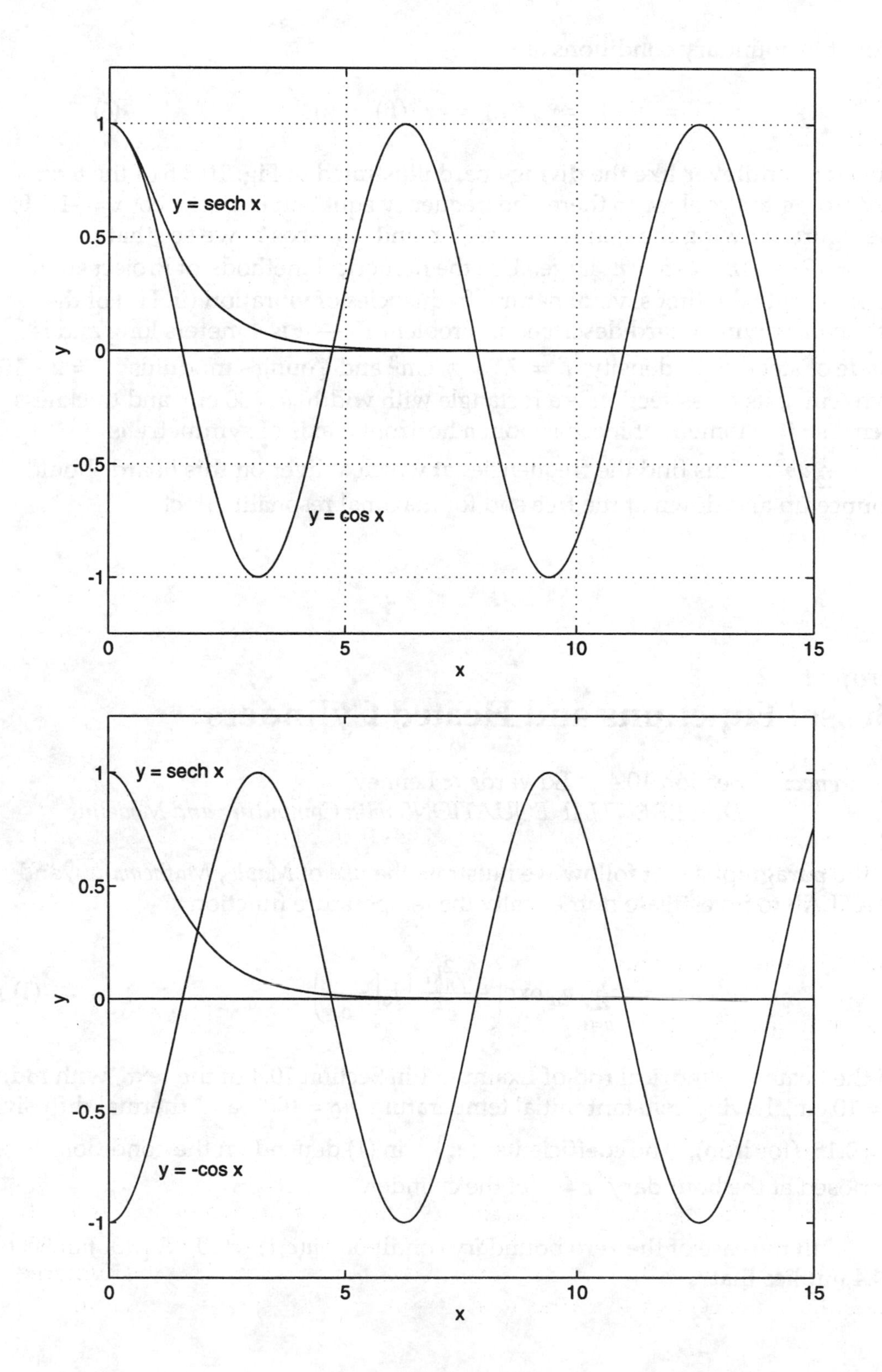

Case 3: Fixed at $x = 0$, free at $x = L$

Now the boundary conditions are

$$y(0) = y'(0) = y''(L) = y'''(L) = 0. \tag{6}$$

This is a **cantilever** like the diving board illustrated in Fig. 10.3.5 of the text. According to Problem 15 there, the frequency equation is $\cosh x \cos x = -1$. From the figure showing the graphs $y = \operatorname{sech} x$ and $y = -\cos x$ we see that $\beta_n \approx (2n-1)\pi/2$ for n large. Use the numerical methods of Project 40 to approximate the first several natural frequencies of vibration (in Hz) of the particular diving board described in Problem 15 — it is 4 meters long and is made of steel (with density $\delta = 7.75$ g/cm^3 and Young's modulus $E = 2 \times 10^{12}$ dyn/cm^2); its cross section is a rectangle with width $a = 30$ cm and thickness $b = 2$ cm, so its moment of inertia about a horizontal axis of symmetry is $I = \frac{1}{12}ab^3$. Thus find the frequencies at which a diver on this board should bounce up and down at the free end for maximal resonant effect.

Project 42
Bessel Functions and Heated Cylinders

Reference: Section 10.4 of Edwards & Penney
DIFFERENTIAL EQUATIONS with Computing and Modeling

In the paragraphs that follow we illustrate the use of *Maple*, *Mathematica*, and MATLAB to investigate numerically the temperature function

$$u(r,t) = \sum_{n=1}^{\infty} a_n \exp\left(-\frac{\gamma_n^2 kt}{c^2}\right) J_0\left(\frac{\gamma_n r}{c}\right) \tag{1}$$

of the heated cylindrical rod of Example 1 in Section 10.4 of the text, with radius $c = 10$ cm, having constant initial temperature $u_0 = 100°$ and thermal diffusivity $k = 0.15$ (for iron). The coefficients $\{a_n\}_1^\infty$ in (1) depend on the condition imposed at the boundary $r = c$ of the cylinder.

In the case of the zero boundary condition $u(c,t) \equiv 0$, Eq. (30) in Section 10.4 implies that

$$a_n = \frac{2u_0}{\gamma_n J_1(\gamma_n)}, \tag{2}$$

where the $\{\gamma_n\}_1^\infty$ are the positive solutions of the equation

$$J_0(x) = 0. \tag{3}$$

Once the first so many of these values have been determined numerically, we can sum the series in (1) so as to calculate values of u and plot $u(r,t)$ versus either r or t.

For your very own cylindrical rod with constant initial temperature $u(r,0) \equiv u_0$ to investigate in the manner we illustrate, let $c = 2p$ and $k = (0.1)q$, where p is the largest and q the smallest nonzero digit of your student I.D. number. For instance, the graph of $u = u(r,120)$ in the first figure below shows how the temperature within the rod after 2 minutes varies with the distance r from its center, and we see that the center-line temperature has already fallen below 60°. The graph of $u = u(0,t)$ for a 5-minute period shown in the next figure indicates that the center-line temperature takes a bit more than 200 seconds to fall to 25°.

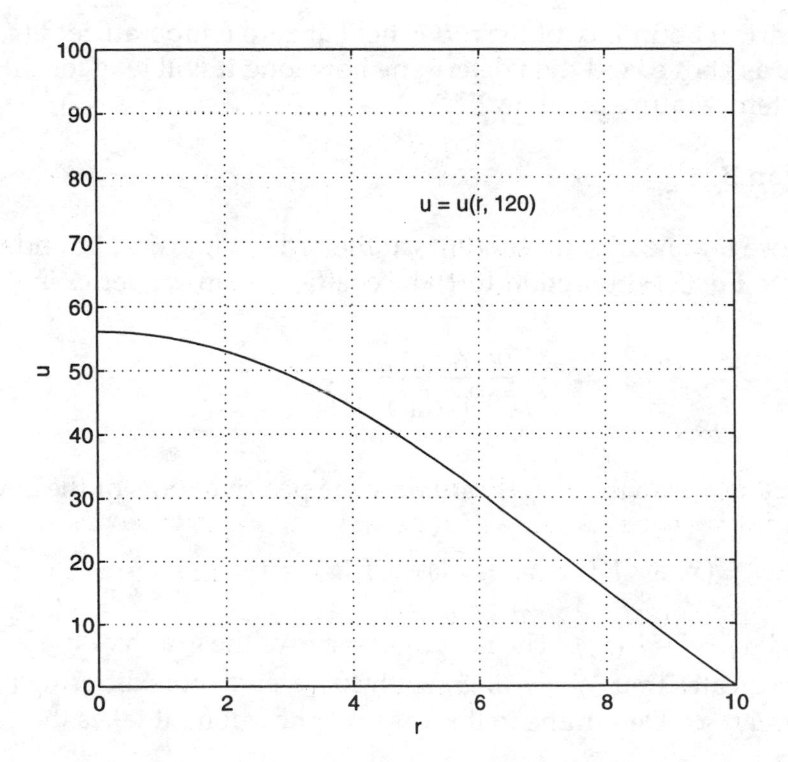

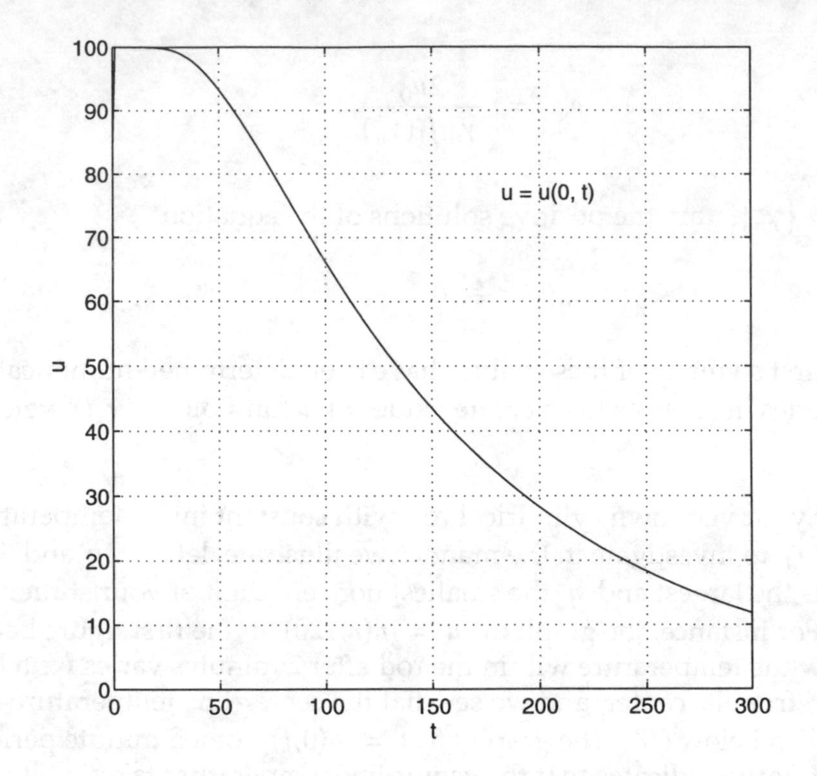

Investigation A

If the cylindrical boundary of the rod is held at zero temperature $u(c,t) = 0$, plot graphs as above, and then determine how long it will take for the rod's center-line temperature to fall to 25°.

Investigation B

Now suppose that heat transfer occurs at the rod's cylindrical boundary, so that (according to Eq. (34) in Section 10.4) the coefficients in the series in (1) are given by

$$a_n = \frac{2u_0\gamma_n J_1(\gamma_n)}{(\gamma_n^2 + h^2)J_0(\gamma_n)^2}. \tag{4}$$

Assume that $h = 1$ so the $\{\gamma_n\}_1^\infty$ are now the positive roots of the equation

$$J_0(x) + x\,J_0'(x) = J_0(x) - x\,J_1(x) = 0 \tag{5}$$

(because $J_0'(x) = -J_1(x)$). The next figure shows the graph of the left-hand side in (5) and indicates that $\gamma_1 \approx 1.25$, with successive roots differing (as usual) by approximately π. Determine in this case *both* how long it takes the centerline temperature to fall to 25° *and* how long it takes the boundary temperature at $r = c$ to fall to 25°.

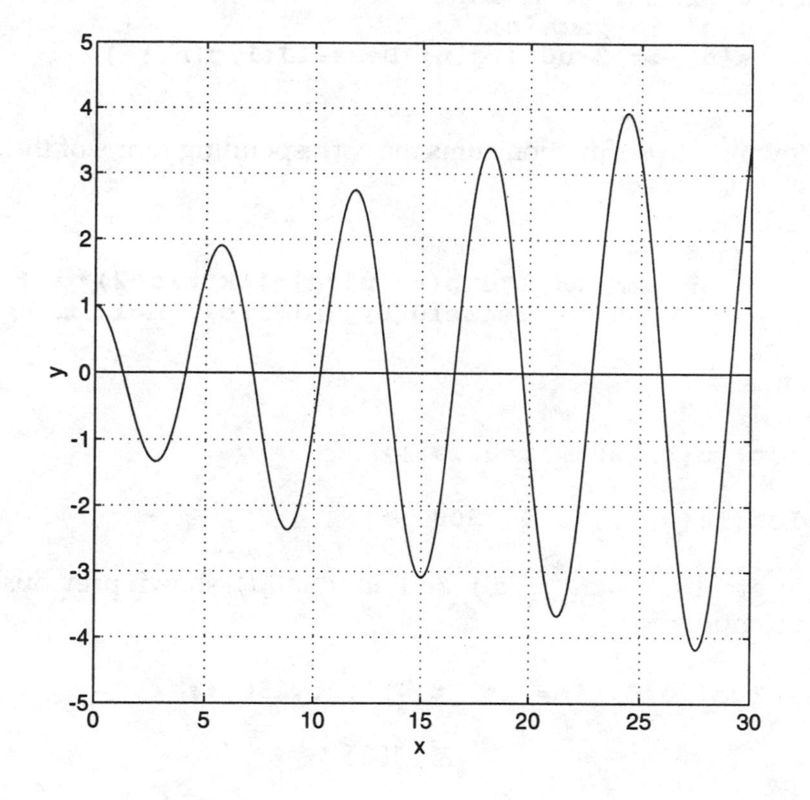

Using *Maple*

Knowing that $\gamma_1 \approx 2.4$ and that successive roots differ approximately by an integral multiple of π, we can approximate the first m values of γ_n by using the *Maple* commands

```
> m := 20:
> gama := array(1..m):        # 'gamma' is reserved for
                              # the gamma function
> for n from 1 to m do
      gama[n] := fsolve( BesselJ(0,x) = 0,
                        x=2+(n-1)*Pi..3+(n-1)*Pi ):
      od:
> seq( gama[n], n=1..m );

{2.40483, 5.52008, 8.65373, 11.7915, 14.9309,
 18.0711, 21.2116, 24.3525, 27.4935, 30.6346,
 33.7758, 36.9171, 40.0584, 43.1998, 46.3412,
 49.4826, 52.6241, 55.7655, 58.9070, 62.0485}
```

Then the first m coefficients in (1) are given by

```
> g := array(1..m):    a := array(1..m):
> c := 10:        u0 := 100:      k := 0.15:
> for n from 1 to m do
        g[n]  := gama[n]:
        a[n]  := 2*u0/( g[n]*BesselJ(1,g[n]) ):
        od:
```

and the following *Maple* function sums the corresponding terms of the series.

```
> n := 'n':
> u := (r,t) ->
           sum( a[n]*exp(-g[n]*g[n]*k*t/c^2)*
                   BesselJ(0,g[n]*r/c), n=1..m );
```

The commands

```
> plot( u(r,120), r=0..c );
```

and

```
> plot( u(0,t), t=0..300 );
```

construct the graphs $u = u(r,120)$ and $u = u(0,t)$ shown previously. Finally, the computation

```
> fsolve( u(0,t) = 25, t = 200..250 );
```

$$214.1025$$

shows that it takes about 214 seconds for the cylinder's center-line temperature to fall to 25°.

Using *Mathematica*

Knowing that $\gamma_1 \approx 2.4$ and that successive roots differ approximately by an integral multiple of π, we can approximate the first m values of γ_n by using the *Mathematica* command

```
m = 20;
solutions =
Table[ FindRoot[ BesselJ[0,x] == 0,
             {x, 2.4 + (n - 1)*Pi} ], {n, 1, m} ];
gamma = x /. solutions

{2.40483, 5.52008, 8.65373, 11.7915, 14.9309,
 18.0711, 21.2116, 24.3525, 27.4935, 30.6346,
 33.7758, 36.9171, 40.0584, 43.1998, 46.3412,
 49.4826, 52.6241, 55.7655, 58.9070, 62.0485}
```

Then the first m coefficients in (1) are given by

```
g = gamma;
c = 10;     u0 = 100;     k = 0.15;
a = 2*u0/(g*BesselJ[1,g]);
```

and the following *Mathematica* function sums the corresponding terms of the series.

```
u[r_,t_] :=
Apply[ Plus,
        a*Exp[-g*g*k*t/c^2]*BesselJ[0,g*r/c] ] // N
```

The commands

```
Plot[ Evaluate[ u[r,120] ], {r, 0,c} ]
```

and

```
Plot[ Evaluate[ u[0,t] ], {t, 0,300} ]
```

construct the graphs $u = u(r,120)$ and $u = u(0,t)$ shown previously. Finally, the computation

```
FindRoot[ u[0,t] == 25, {t, 200} ]

{t -> 214.013}
```

shows that it takes about 214 seconds for the cylinder's center-line temperature to fall to 25°.

Using MATLAB

First we define the equation $J_0(x) = 0$ that we need to solve in the form

```
function   y = j0(x)
y = besselj(0,x);
```

Then knowing that $\gamma_1 \sim 2.4$ and that successive roots differ approximately by an integral multiple of π, we can approximate the first m values of γ_n by using the MATLAB commands

```
» m = 20;
» for  n = 1 : m
        g(n) = fzero('j0', 2.4 + (n-1)*pi );
        end

»reshape(g,5,4)'
ans =
```

```
   2.4048     5.5201     8.6537    11.7915    14.9309
  18.0711    21.2116    24.3525    27.4935    30.6346
  33.7758    36.9171    40.0584    43.1998    46.3412
  49.4826    52.6241    55.7655    58.9070    65.1900
```

Ever alert, we notice that the `g(20)` value is incorrect, and proceed to correct it.

```
» g(20) = fzero('j0', g(19) + pi );
» g(20)
ans =   62.0485
```

Now that these γ_n values are available, we can now edit the function `u.m` of Project 40 so as to calculate the $p \times q$ matrix `u(r,t,g)` of temperatures corresponding to a q-vector `r` of radii and a p-vector `t` of times:

```
function    u = u(r,t,g)
%  r = row vector of  q  radii of cylinder
%  t = row vector of  p  times
%  u = p x q  matrix of temperatures

c = 10;                      % radius of cylinder
u0 = 100;                    % initial temp of cylinder
k = 0.15;                    % thermal diffusivity
m = length(g);               % no of terms to use

a = 2*u0./(g.*besselj(1,g));       % coefficients in (1)
coeffs = diag(a);                  % m x m  diagonal
                                   % matrix of coeffs
exps = exp(-(g'.*g')*k*t/c^2);     %  m x p  matrix
bessels = besselj(0,g'*r/c);       %  m x q  matrix
u = exps'*coeffs*bessels;          %  p x q  matrix
```

The commands

```
» r = 0 : 0.05 : 10;
» plot( r, u(r,120,g) )
```

and

```
» t = 0 : 300;
» plot( t, u(0,t,g) )
```

construct the graphs $u = u(r,120)$ and $u = u(0,t)$ shown previously. Finally, if we define the temperature function

```
function   y = u0(t)
g = ...
  [ 2.4048     5.5201     8.6537    11.7915    14.9309 ...
   18.0711    21.2116    24.3525    27.4935    30.6346 ...
   33.7758    36.9171    40.0584    43.1998    46.3412 ...
   49.4826    52.6241    55.7655    58.9070    65.1900 ];
y = u(0,t,g) - 25;
```

(simply copying the γ_n values from above) then the computation

```
» fzero('u0',200)
ans =
      214.1071
```

shows that it takes about 214 seconds for the cylinder's center-line temperature to fall to 25°.

Project 43

Diffusion of Heat in a Rectangular Plate

Reference: Section 10.5 of Edwards & Penney *DIFFERENTIAL EQUATIONS with Computing and Modeling*

Example 1 in Section 10.5 of the text involves heat flow in a thin rectangular plate that occupies the plane region $0 \le x \le a,\ 0 \le y \le b$. It has insulated faces and its four edges are held at temperature zero. If it has the given initial temperature function $f(x,y)$, then its temperature function $u(x,y,t)$ satisfies the boundary value problem consisting of the 2-dimensional heat equation

$$\frac{\partial u}{\partial t} = k\,\nabla^2 u = k\left(\frac{\partial^2 u}{\partial x^2} + \frac{\partial^2 u}{\partial y^2}\right) \tag{1}$$

and the boundary conditions

$$u(0,y,t) = u(a,y,t) = u(x,0,t) = u(x,b,t) = 0, \tag{2}$$

$$u(x,y,0) = f(x,y). \tag{3}$$

If $f(x,y) \equiv u_0$, a constant, then Problem 1 in the text yields the double Fourier series solution

$$u(x,y,t) = \frac{16u_0}{\pi^2} \sum_{m\ odd} \sum_{n\ odd} \frac{\exp(-\gamma_{mn}^2 kt)}{mn} \sin\frac{m\pi x}{a} \sin\frac{n\pi y}{b} \tag{4}$$

where

$$\gamma_{mn}^2 = \left(\frac{m^2}{a^2}+\frac{n^2}{b^2}\right)\pi^2. \qquad (5)$$

Here we will take $a = b = \pi$, $u_0 = 100°$, and $k = 0.5$ for illustrative purposes.

Using *Maple*

We can define the temperature function $u(x,y,t)$ in (4) using the commands

```
> k :=0.5:     u0 := 100:

> c := (m,n) -> sqrt(m^2 + n^2):

> exps := (m,n,t) -> exp(-c(m,n)^2*k*t)/(m*n):

> N := 6:       # use first N odd m and n indices

> u := (x,y,t) ->
       (16*u0/Pi^2)*sum(sum(
                              exps(2*m-1,2*n-1,t)*
                              sin((2*m-1)*x)*sin((2*n-1)*y),
                                'm'=1..N),'n'=1..N):
```

For instance, at time $t = 1$ the temperature at the center point $(\pi/2, \pi/2)$ of our $\pi\times\pi$ square plate is given by

```
> evalf(u(Pi/2,Pi/2,1));
```

58.912527

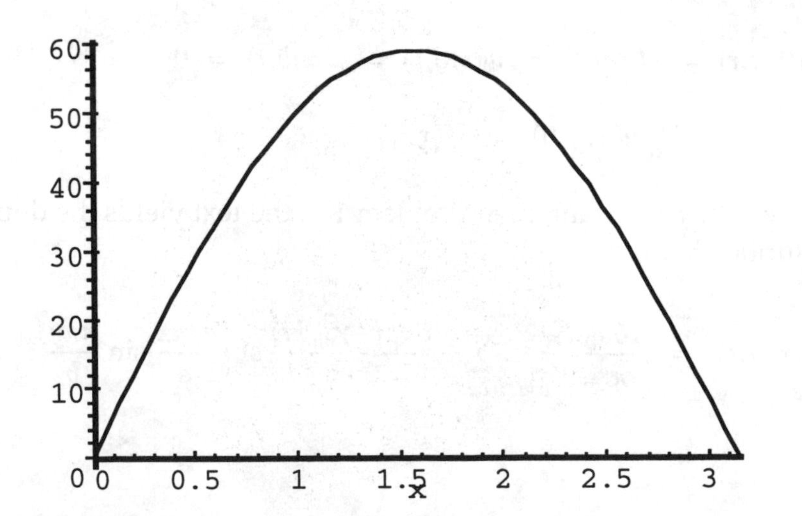

With $N = 6$, we have summed only the terms for $m, n = 1, 3, 5, 7, 9, 11$ in (4), but you can check that with $N = 12$ (for instance) we get the same numerical result accurate to 6 decimal places as shown. The command

```
> plot( u(x,Pi/2,1), x=0..Pi );
```

generates the graph $u = u(x, \pi/2, 1)$ which shows the expected symmetry about the center point, where the maximum temperature apparently is attained.

Using *Mathematica*

We can define the temperature function $u(x,y,t)$ in (4) using the commands

```
k = 0.5;         u0 = 100;

exps[m_,n_,t_]  := Exp[-(m^2 + n^2) k t]/(m*n)

M = 15;                    (* ODD maximal index *)

u[x_,y_,t_] :=
    (16*u0/Pi^2) Sum[ Sum[ exps[m,n,t] Sin[m x] Sin[n y],
                   {m,1,M,2} ], {n,1,M,2} ]
```

For instance, at time $t = 1$ the temperature at the center point $(\pi/2, \pi/2)$ of

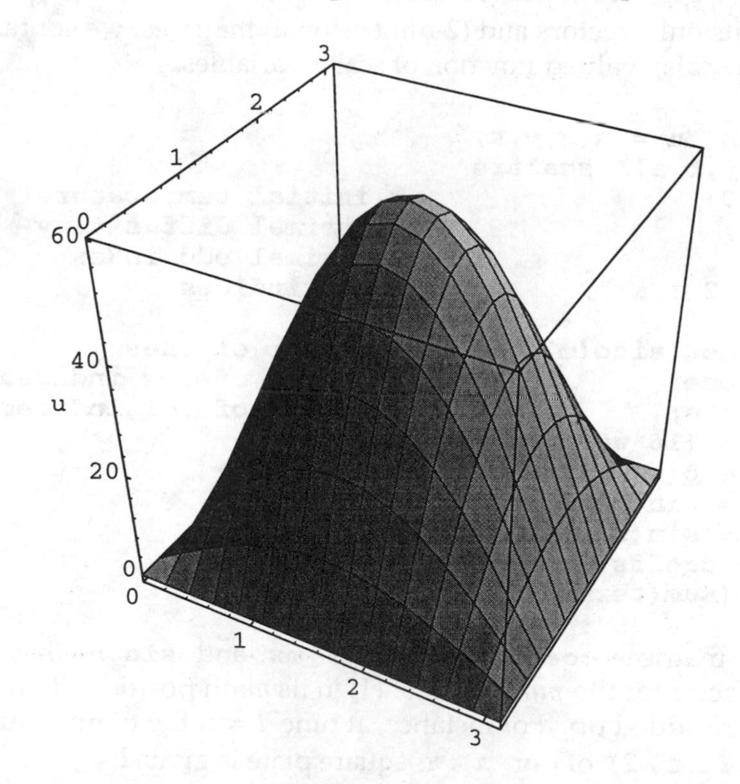

our $\pi \times \pi$ square plate is given by

```
u[ Pi/2, Pi/2, 1 ] // N
```

```
58.9125
```

With $N = 15$, we have summed only the terms for $m, n = 1, 3, 5, 7, 9, 11, 13, 15$ in (4), but you can check that with $N = 25$ (for instance) we get the same numerical result accurate to 6 significant digits as shown. The command

```
Plot3D[ Evaluate[ u[x,y,1] ], {x,0,Pi}, {y,0,Pi},
        BoxRatios -> {1,1,1},
        AxesLabel -> {"","","u"} ]
```

generates the graph $u = u(x,y,1)$, which shows the expected symmetry about the center point of the plate, where the maximum temperature at time $t = 1$ evidently is attained.

Using MATLAB

For convenience in plotting the temperature u as a function of any one or two of the variables x, y, or t, it would be best to define a function **u(x,y,t)** that accepts a p-vector **x**, a q-vector **y**, and an r-vector **t** and returns a 3-dimensional $p \times q \times r$ array of u-values. However, since MATLAB (prior to version 5) permits only vectors and (2-dimensional) matrices, we content ourselves with a scalar-valued function of scalar variables.

```
function  u = u(x,y,t)
%  u,x,y,t all scalars
u0 = 100;                    % initial temperature
k = 0.5;                     % thermal diffusivity
N = 25;                      % maximal odd index
m = 1 : 2 : N;               % odd indices
n = m;
one = ones(size(m));         % vector of ones
m = m'*one;                  % matrix of row indices
n = one'*n;                  % matrix of col indices
coeffs = (16*u0/pi^2) ./(m.*n);
exps = exp(-(m.^2+n.^2)*k*t);
sin_mx = sin(m*x);
sin_ny = sin(n*y);
terms = coeffs.*exps.*sin_mx.*sin_ny;
u = sum(sum(terms));
```

Here each of the matrices **coeffs**, **exps**, **sin_mx**, and **sin_ny** has the corresponding factor for the *mn*th term of (1) in its *mn*th position. In the last line all these terms are added up. For instance, at time $t = 1$ the temperature at the center point $(\pi / 2, \pi / 2)$ of our $\pi \times \pi$ square plate is given by

```
» u(pi/2,pi/2,1)
ans =
    58.9125
```

When we plot $u(\pi/2, \pi/2, t)$ for $0 \le t \le 5$ with the commands

```
» t = 0 : 0.05 : 5;
» for n = 1 : length(t)
      uc(n) = u(pi/2,pi/2,t(n));
      end
» plot(t,uc)
```

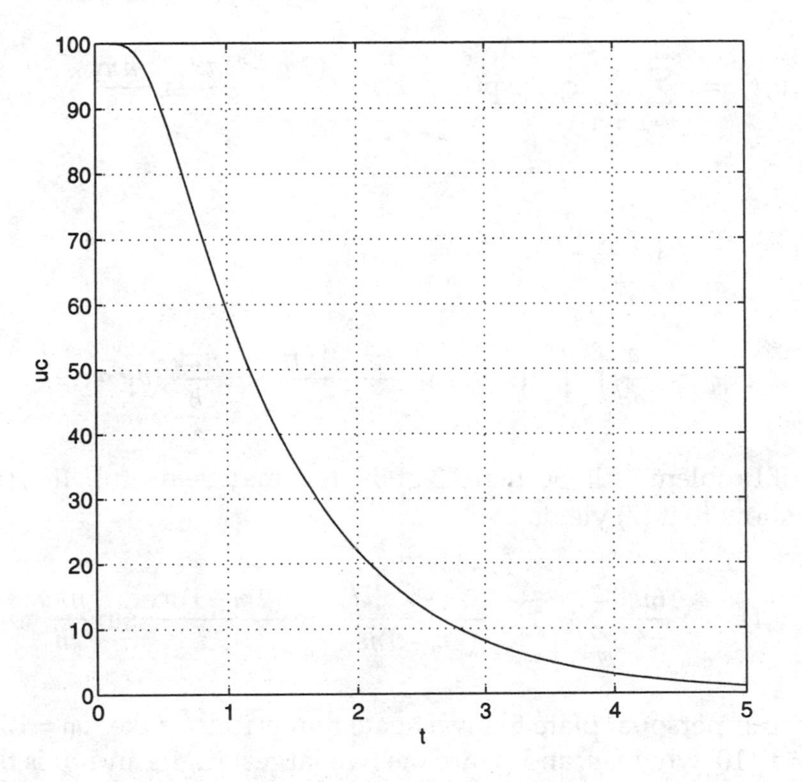

we see that the plate's center-point temperature falls to 20° in just over 2 seconds. In order to use **fzero** to approximate this time accurately, we express the difference between 20 and u at $(\pi/2, \pi/2)$ as a function of the single variable t:

```
function   v = uc(t)
v = u(pi/2,pi/2,t) - 20;
```

Then the computation

```
» fzero('uc',2)
ans = 2.0924
```

shows that $t \approx 2.0924$ when the center-point temperature hits 20°.

Investigation

Suppose the three edges $x = 0,\ y = 0,$ and $y = b$ of the plate of Example 1 are held at temperature zero, but the fourth edge $x = a$ is insulated, with corresponding boundary conditions

$$u(0,y,t) = u(x,0,t) = u(x,b,t) = u_x(a,y,t) = 0. \tag{6}$$

If the plate's initial temperature is $u(x,y,0) = f(x,y)$, show that its temperature function is given by

$$u(x,y,t) = \sum_{m=1}^{\infty}\sum_{n=1}^{\infty} c_{mn} \exp(-\gamma_{mn}^2 kt) \sin\frac{(2m-1)\pi x}{2a} \sin\frac{n\pi y}{b} \tag{7}$$

where

$$\left(\frac{\gamma_{mn}}{\pi}\right)^2 = \left(\frac{2m-1}{a}\right)^2 + \left(\frac{n}{b}\right)^2 \tag{8}$$

and

$$c_{mn} = \frac{4}{ab}\int_0^a\int_0^b f(x,y) \sin\frac{(2m-1)\pi x}{2a} \sin\frac{n\pi y}{b}\, dy\, dx. \tag{9}$$

The result of Problem 21 in Section 9.3 of the text may be useful. If $f(x,y) \equiv u_0$ (constant), show that (7) yields

$$u(x,y,t) = \frac{16u_0}{\pi^2}\sum_{m=1}^{\infty}\sum_{n\ odd} \frac{\exp(-\gamma_{mn}^2 kt)}{(2m-1)n} \sin\frac{(2m-1)\pi x}{2a} \sin\frac{n\pi y}{b}. \tag{10}$$

For your personal plate to investigate numerically, take $u_0 = 100$, $a = 10p$, $b = 10q$, $k = r/10$ where p and q are the two largest digits and r is the smallest nonzero digit in you student I.D. number. Plot $u = u(x,y,t)$ as a function of x and y for typical values of t to verify that each such graph is symmetric with respect to the midline $y = b/2$ of the plate, so it follows (why?) that the maximum temperature in the plate at time t occurs at some point of this midline. Then determine

- How long it takes for the maximum temperature on the edge $x = a$ to fall to 20°.
- What is then the maximum interior temperature in the plate.

Project 44
Rectangular Membrane Vibrations

Reference: Section 10.5 of Edwards & Penney
DIFFERENTIAL EQUATIONS with Computing and Modeling

Here we investigate the vibrations of a flexible membrane whose equilibrium position is the rectangle $0 \le x \le a,\ 0 \le y \le b$. Suppose it is released from rest with given initial displacement, and thereafter its four edges are held fixed. Then (under the usual assumptions) its displacement function $u(x,y,t)$ satisfies the boundary value problem

$$\frac{\partial^2 u}{\partial t^2} = c^2\left(\frac{\partial^2 u}{\partial x^2}+\frac{\partial^2 u}{\partial y^2}\right) \qquad \left(c^2 = T/\rho\right) \tag{1}$$

$$u(0,y,t) = u(a,y,t) = u(x,0,t) = u(x,b,t) = 0 \tag{2}$$

$$u(x,y,0) = f(x,y)$$
$$u_t(x,y,0) = 0. \tag{3}$$

According to Problem 3 in Section 10.5 of the text, the solution is given by

$$u(x,y,t) = \sum_{m=1}^{\infty}\sum_{n=1}^{\infty} c_{mn} \sin\frac{m\pi x}{a} \sin\frac{n\pi y}{b} \cos \gamma_{mn}ct \tag{4}$$

where the coefficients are defined by

$$c_{mn} = \frac{4}{ab}\int_0^a\int_0^b f(x,y) \sin\frac{m\pi x}{a} \sin\frac{n\pi y}{b}\, dy\, dx. \tag{5}$$

The mnth term in (4) corresponds to the membrane's mnth natural mode of oscillation with displacement function

$$u_{mn}(x,y,t) = \sin\frac{m\pi x}{a} \sin\frac{n\pi y}{b} \cos \gamma_{mn}ct \tag{6}$$

with circular frequency $\omega_{mn} = \gamma_{mn}c$ where

$$\gamma_{mn}^2 = \left(\frac{m^2}{a^2}+\frac{n^2}{b^2}\right)\pi^2. \tag{7}$$

The *mn*th inital position function

$$u_{mn}(x,y) = \sin\frac{m\pi x}{a}\sin\frac{n\pi y}{b} \tag{8}$$

is the rectangular membrane's ***mn*th eigenfunction**.

Investigation

For simplicity take $a = b = c = 1$ and plot some eigenfunctions with small values of m and n in (8). Then plot linear combinations of several eigenfunctions to see some of the more interesting possible initial shapes of a vibrating membrane. For example, the figure below shows the graph of the initial position function $u(x,y) = u_{11}(x,y) - 3\,u_{22}(x,y)$ generated by the *Mathematica* commands

```
u = Sin[x] Sin[y] - 3 Sin[2x] Sin[2y];
Plot3D[ Evaluate[u], {x,0,Pi}, {y,0,Pi},
        PlotPoints -> {20,20},
        ViewPoint -> {-1.5,3,0.5} ]
```

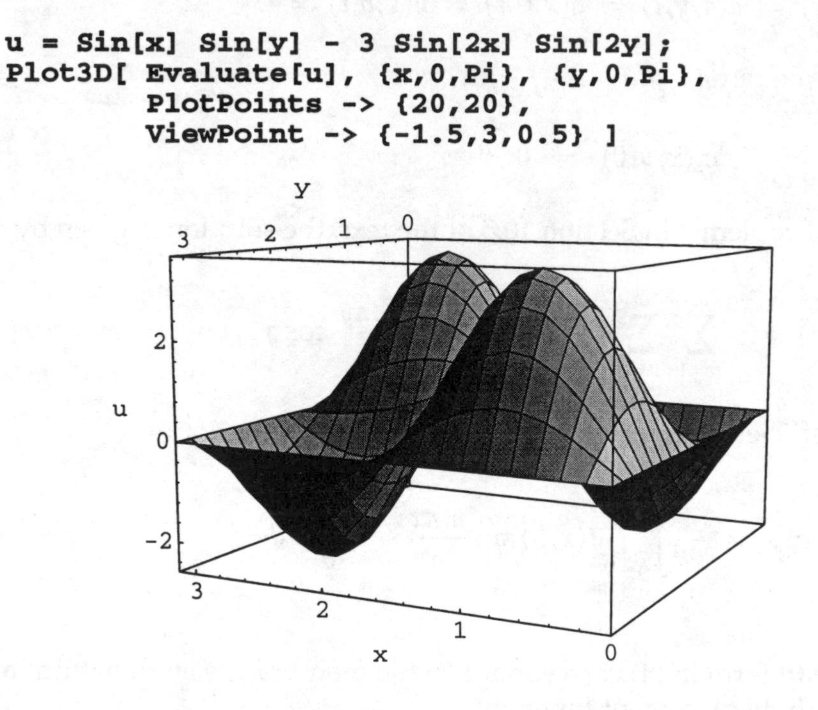

The *Maple* commands to generate this figure would be very similar; the corresponding MATLAB code is

```
x = 0 : pi/20 : pi;      y = x;
[x,y] = meshgrid(x,y);
u = sin(x).*sin(y) - 3*sin(2*x).*sin(2*y);
surf(x,y,u)
```

Maple, *Mathematica*, and MATLAB all have the capability to animate a sequence of snapshots of a vibrating membrane so as to show a "movie" illustrating its motion. For instance the *Maple* commands

```
> with(plots):

> u  := (x,y) -> sin(x)*sin(2*y) + sin(2*x)*sin(y):

> w := sqrt(5):       # circular frequency
> p := 2*Pi/w:        # period of oscillation

> animate3d( u(x,y)*cos(w*t),
             x=0..Pi, y=0..Pi, t=0..p,
             frames=12,
             style = patch );
```

produce a 12-frame movie showing one complete oscillation of the membrane with initial position function

$$u(x,y) = \sin x \sin 2y + \sin 2x \sin y \tag{9}$$

and circular frequency $\omega = \sqrt{5}$. The *Mathematica* commands

```
u = Sin[x] Sin[2y] + Sin[2x] Sin[y];

w = Sqrt[5];     (* circular frequency    *)
P = 2 Pi/w;      (* period of oscillation *)

frame = Table[ Plot3D[ Evaluate[u Cos[w t]],
                   {x,0,Pi}, {y,0,Pi},
                    PlotRange -> {-1.5, 1.5},
                    BoxRatios -> {3,3,2},
                    ViewPoint->{-1.5, 2.8, 0.75} ],
                    {t,0,P/2, P/20 } ];
```

produce a movie of a half-oscillation which (with the **Animate Graphics** selection) can be played back-and-forth to show successive oscillations continuously. The command

```
Show[ GraphicsArray[ {{frame[[1]],  frame[[ 3]]},
                      {frame[[5]],  frame[[ 7]]},
                      {frame[[9]],  frame[[11]]}} ]]
```

displays the array of successive snapshots shown on the next page.

Experiment in this way with linear combinations of two, three, or more membrane eigenfunctions of the form

$$u_{mn}(x,y,t) = \sin mx \sin ny \cos \omega_{mn} t \tag{10}$$

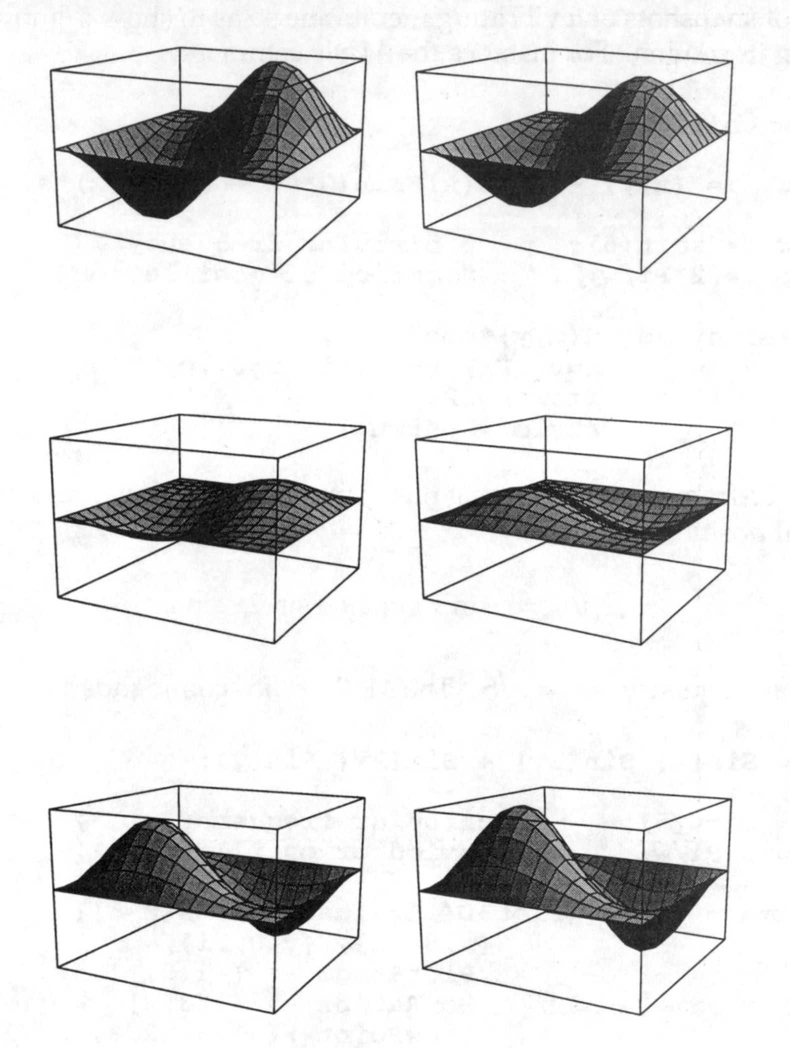

where $\omega_{mn} = \sqrt{m^2+n^2}$. Vary the coefficients so as to produce a visually attractive movie of sufficient complexity to be interesting. If you're using MATLAB, see the commands illustrated in the investigation that follows.

The Plucked Square Membrane

Suppose the square membrane $0 \leq x, y \leq \pi$ is plucked at its center point and set in motion from rest with the initial position function

$$u(x,y,0) = f(x,y) = \min(x, y, \pi - x, \pi - y) \tag{11}$$

whose graph over the square $0 \le x, y \le \pi$ looks like a square tent or pyramid with height $\pi/2$ at its center. Thus the "tent function" $f(x,y)$ is the 2-dimensional analogue of the familiar 1-dimensional triangle function that describes the initial position of a plucked string. It can be defined "piecewise" as indicated in the following figure.

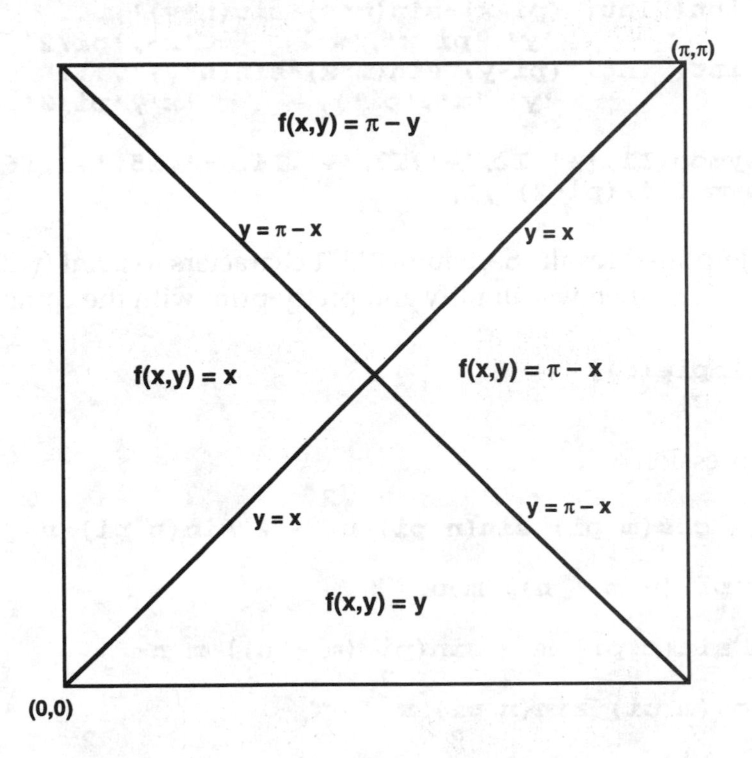

This diagram indicates how to subdivide the integral in (5), with $a = b = \pi$ in order to calculate the coefficients $\{c_{mn}\}$ in (4). We write

$$f(x,y) = \begin{cases} y & \text{if } 0 < x < \pi/2,\ 0 < y < x, \\ x & \text{if } 0 < x < \pi/2,\ x < y < \pi - x, \\ \pi - y & \text{if } 0 < x < \pi/2,\ p - x < y < \pi, \\ y & \text{if } \pi/2 < x < \pi,\ 0 < y < \pi - x, \\ \pi - x & \text{if } \pi/2 < x < \pi,\ \pi - x < y < x, \\ \pi - y & \text{if } \pi/2 < x < \pi,\ x < y < \pi. \end{cases} \tag{12}$$

Then the coefficient integral is the sum of six integrals corresponding to the six "parts" of the defintion in (12). To evaluate these integrals symbolically, we enter the MATLAB (Symbolic Math Toolbox) commands

```
I1 = int( int('y*sin(m*x)*sin(n*y)',...
              'y',0,'x'),          'x',0,'pi/2');
```

```
I2 = int( int('x*sin(m*x)*sin(n*y)',...
                    'y','x','pi-x'),        'x',0,'pi/2');
I3 = int( int('(pi-y)*sin(m*x)*sin(n*y)',...
                    'y','pi-x','pi'),       'x',0,'pi/2');
I4 = int( int('y*sin(m*x)*sin(n*y)',...
                    'y',0,'pi-x'),          'x','pi/2','pi');
I5 = int( int('(pi-x)*sin(m*x)*sin(n*y)',...
                    'y','pi-x','x'),        'x','pi/2','pi');
I6 = int( int('(pi-y)*sin(m*x)*sin(n*y)',...
                    'y','x','pi'),          'x','pi/2','pi');

c = symop(I1,'+',I2,'+',I3,'+',I4,'+',I5,'+',I6);
c = symm('4/(pi^2)',I);
```

The final unsimplified result **b** requires 4371 characters to print (with no spaces). However, when we simplify and pretty-print with the commands

```
c = simple(c);
pretty(c)
```

we get the expression

```
                                2                    2
[ 2 (2 cos(m pi) sin(n pi) n   - 2 sin(n pi) n

+ sin(pi (- m + n)) m n

+ 2 n sin(m pi) m - sin(pi (m + n)) m n
                          2
- 2 cos(m pi) sin(n pi) m
                         2                        2
+ sin(pi (- m + n)) m   + sin(pi (m + n)) m )

      /                          2    2
     /   ((m + n) (- m + n) n   m pi  ) ]
    /
```

for the coefficient c_{mn}. We see (from the denominator) that MATLAB is assuming m and n not equal, in which case $c_{mn} = 0$ because of the sine factors.

To calculate the non-zero "diagonal coefficients" in the Fourier series (4), we repeat the computation above with $m = n$ from the beginning. The result is

$$c_{nn} = \frac{2\left[1-(-1)^n\right]}{\pi n^2} = \begin{cases} 4/\pi n^2 & \text{for } n \text{ odd,} \\ 0 & \text{for } n \text{ even.} \end{cases} \tag{13}$$

Thus the Fourier series of the tent function $f(x,y)$ defined in (11) and (12) is

$$f(x,y) = \frac{4}{\pi} \sum_{n\ odd}^{\infty} \frac{\sin nx \sin ny}{n^2}. \tag{14}$$

It follows that the solution of our original vibrating membrane problem with initial position function $f(x, y)$ is given by

$$u(x,y,t) = \frac{4}{\pi} \sum_{n\ odd}^{\infty} \frac{\sin nx \sin ny \cos nt\sqrt{2}}{n^2}. \tag{15}$$

We invite you check out the computation outlined here using either *Maple*, *Mathematica*, or MATLAB. Try unequal values of a and b to see whether you still get a "diagonal" series as in (15).

Is it clear — because (15) contains no terms with $m \neq n$ — that the function $u(x,y,t)$ is periodic (in t) with period $P = \pi\sqrt{2}$? The fact that the tent function in (15) thus yields a "musical" vibration of a square membrane was first pointed out to us by John Polking.

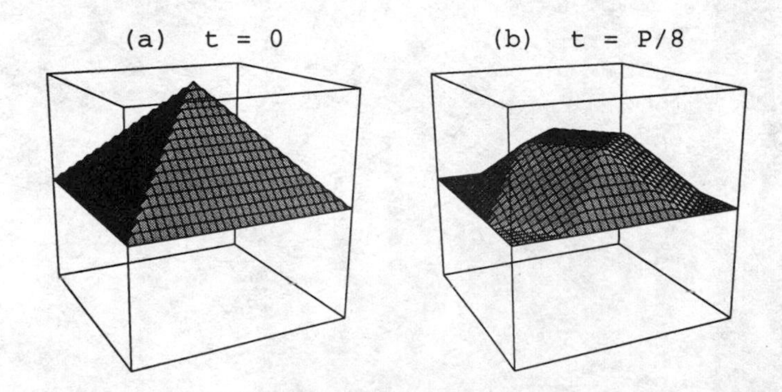

The following *Mathematica* commands generate the two snapshots shown above.

```
n = 25;        (* no of terms used   *)

u[x_,y_,t_] :=
(4/Pi) Sum[ Sin[k x] Sin[k y] Cos[k t Sqrt[2]]/(k*k),
            {k,1,2n - 1, 2} ]

P = Pi Sqrt[2];   (* Period of oscillation   *)

figA =
Plot3D[ Evaluate[u[x,y,0]], {x,0,Pi}, {y,0,Pi},
                    PlotPoints -> {31,31} ];
```

```
figB =
Plot3D[ Evaluate[u[x,y,P/8]], {x,0,Pi}, {y,0,Pi},
                    PlotPoints -> {31,31}];

Show[ GraphicsArray[ {figA, figB} ]];
```

The MATLAB script below generates a movie showing the vibrating membrane with tent-shaped initial position. With a Macintosh system the **qtwrite** command can be used to save this movie in QuickTime format for later viewing or demonstration, or even for incorporation in a word-processing document. Thus, if you are using this manual in electronic format you need only double-click on the movie button in the lower left corner below to view this vibrating membrane in full motion and living color.

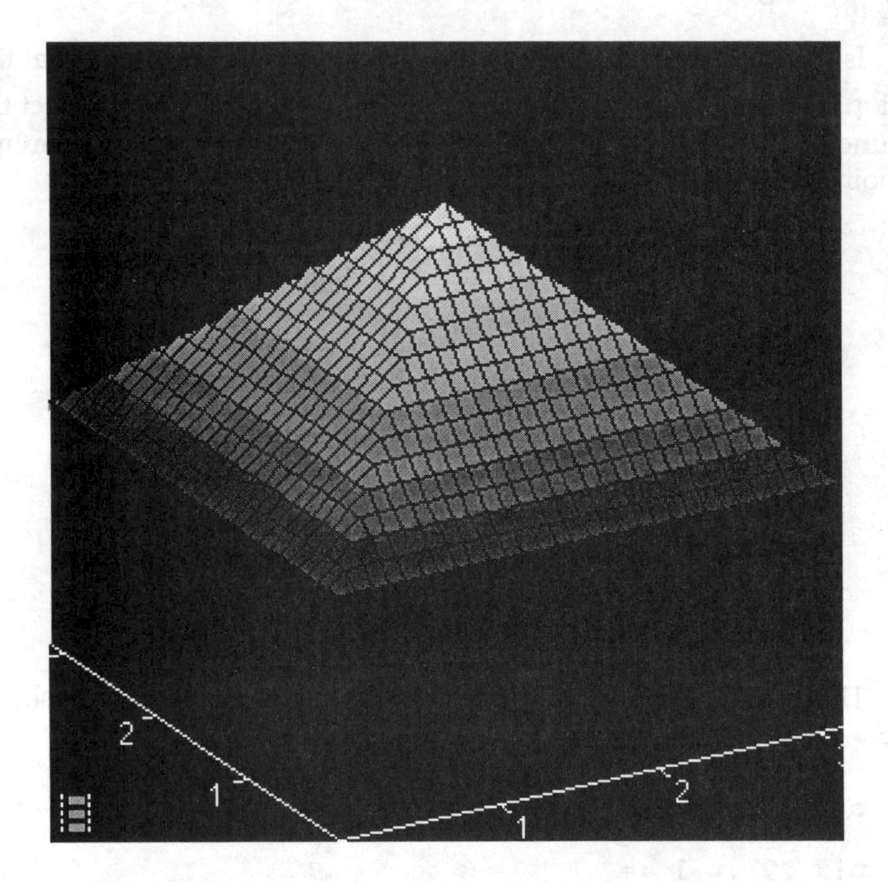


```
% tentmovi.m

x = 0 : pi/30 : pi;
y = x;
[xx,yy] = meshgrid(x,y);
```

```
q = 20;     % no of frames

newplot
axis('square')
Mtent = moviein(q+1);

for  j = 0 : q
     t = pi*j/(q*sqrt(2));
     zz = ztent(x,y,t);
     surf(xx,yy,zz)
     axis('square')
     colormap(hot)
     caxis([-pi/2 pi/2])
     axis([0 pi 0 pi -pi/2 pi/2]);
     view([-30  25])
     Mtent(:,j+1) = getframe;
     end

movie(Mtent,-10)
```

Project 45
Circular Membrane Vibrations

Reference: Sections 10.5 and 10.5 of Edwards & Penney *DIFFERENTIAL EQUATIONS with Computing and Modeling*

In problems involving regions that enjoy circular symmetry about the origin in the plane (or the vertical z-axis in space), the use of polar (or cylindrical) coordinates is advantageous. In Section 9.7 of the text we discussed the expression of the 2-dimensional Laplacian

$$\nabla^2 u = \frac{\partial^2 u}{\partial r^2} + \frac{1}{r}\frac{\partial u}{\partial r} + \frac{1}{r^2}\frac{\partial^2 u}{\partial \theta^2} \tag{1}$$

in terms of the familiar plane polar coordinates (r,θ) for which $x = r\cos\theta$ and $y = r\sin\theta$. If $u(r,\theta,t)$ denotes the vertical displacement at time t of the point (r,θ) of a vibrating circular membrane of radius c, then the 2-dimensional wave equation takes the polar coordinate form

$$\frac{\partial^2 u}{\partial t^2} = a^2\nabla^2 u = a^2\left(\frac{\partial^2 u}{\partial r^2} + \frac{1}{r}\frac{\partial u}{\partial r} + \frac{1}{r^2}\frac{\partial^2 u}{\partial \theta^2}\right). \tag{2}$$

where $a^2 = T/\rho$ in terms of the membrane's tension T and density ρ (per unit area). If the membrane is released from rest with given initial position function $f(r,\theta)$ at time $t = 0$ and thereafter its boundary is held fixed, then the membrane's displacement function $u(r,\theta,t)$ satisfies both (2) and the boundary conditions

$$u(c,\theta,t) = 0, \qquad \text{(fixed boundary)} \tag{3}$$

$$u(r,\theta,0) = f(r,\theta), \qquad \text{(given initial displacement)} \tag{4}$$

$$u_t(r,\theta,0) = 0. \qquad \text{(zero initial velocity)} \tag{5}$$

Fill in the details in the solution that is outlined as follows. Show first that the substitution

$$u(r,\theta,t) = R(r)\,\Theta(\theta)\,T(t) \tag{6}$$

in the wave equation (2) yields the separation of variables

$$\frac{T''}{a^2T} = \frac{R'' + \frac{1}{r}R'}{R} + \frac{\Theta''}{r^2\Theta} = -\alpha^2 \qquad \text{(constant).} \tag{7}$$

Then

$$T'' + \alpha^2 a^2 T = 0, \qquad T'(0) = 0 \tag{8}$$

implies that (to within a constant multiple)

$$T(t) = \cos \alpha a t. \tag{9}$$

Next, the right equality in (43) yields the equation

$$\frac{r^2R'' + rR'}{R} + \alpha^2 r^2 + \frac{\Theta''}{\Theta} = 0 \tag{10}$$

from which it follows that

$$\frac{\Theta''}{\Theta} = -\beta^2 \qquad \text{(constant).} \tag{11}$$

In order that a solution $\Theta(\theta)$ of $\Theta'' + \beta^2\Theta = 0$ have the necessary 2π-periodicity, the parameter β must be an integer, so we have the θ-solutions

$$\Theta_n(\theta) = \begin{cases} \cos n\theta \\ \sin n\theta \end{cases} \tag{12}$$

for $n = 0, 1, 2, 3, \ldots\ldots$.

Substitution of $\Theta''/\Theta = -n^2$ in (10) now yields the parametric Bessel equation

$$r^2R'' + rR' + (\alpha^2r^2 - n^2)R = 0 \tag{13}$$

of order n, with bounded solution $R(r) = J_n(\alpha r)$. Since the zero boundary condition (3) yields $J_n(\alpha c) = 0$, case 1 in the table of Fig. 10.4.2 of the text yields the r-eigenfunctions

$$R_{mn}(r) = J_n\left(\frac{\gamma_{mn}r}{c}\right) \qquad (m = 1, 2, 3, \ldots;\ n = 0, 1, 2, \ldots) \tag{14}$$

where γ_{mn} denotes the mth positive solution of the equation $J_n(x) = 0$. Finally, substitution of $\alpha_{mn} = \gamma_{mn}/c$ in (9) yields the t-function

$$T_{mn}(t) = \cos\frac{\gamma_{mn}at}{c}. \tag{15}$$

Combining (12), (14), and (15) we see that our boundary value problem for the circular membrane released from rest has the formal series solution

$$u(r,\theta,t) = \sum_{m=1}^{\infty}\sum_{n=0}^{\infty} J_n\left(\frac{\gamma_{mn}r}{c}\right)(a_{mn}\cos n\theta + b_{mn}\sin n\theta)\cos\frac{\gamma_{mn}at}{c}. \tag{16}$$

Thus the vibrating circular membrane's typical natural mode of oscillation with zero initial velocity is of the form

$$u_{mn}(r,\theta,t) = J_n\left(\frac{\gamma_{mn}r}{c}\right)\cos n\theta\cos\frac{\gamma_{mn}at}{c} \tag{17}$$

or the analogous form with $\sin n\theta$ instead of $\cos n\theta$. In this mode the membrane vibrates with $m-1$ fixed *nodal circles* (in addition to its boundary $r = c$) with radii $r_{jn} = \gamma_{jn}c/\gamma_{mn}$ for $j = 1, 2, \ldots\ldots, m-1$. It also has $2n$ fixed *nodal radii* spaced at angles of π/n starting with $\theta = \pi/2n$. The following figure shows some typical configurations of these nodal circles and radii, which divide the circle into annular sectors that move alternately up and down as the membrane vibrates.

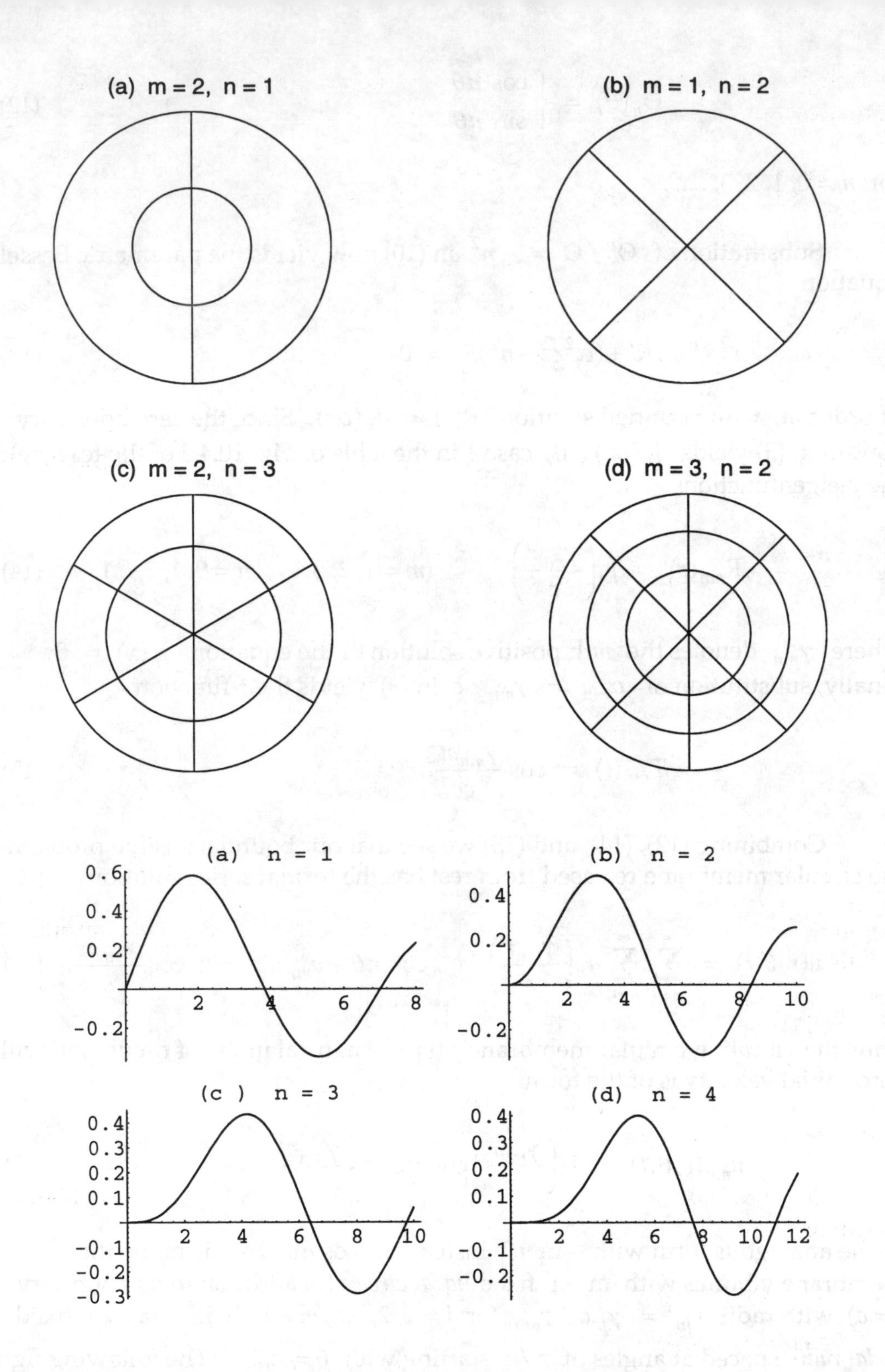

The figure above shows the graphs of $y = J_n(x)$ for $n = 1, 2, 3, 4$. We see that $n + 3$ is (at least for small n) a rough but reasonable initial estimate of the first positive zero γ_{1n} of the equation $J_n(x) = 0$. This observation motivates the *Mathematica* commands

```
inits =
Table[ FindRoot[ BesselJ[n,x] == 0, {x, n+3} ],
       {n, 1,5} ];
g1 = x /. inits

{3.83171, 5.13562, 6.38016, 7.58834, 8.77148}}
```

that accurately approximate these inital zeros.

Now recall that the gap between successive zeros of $J_n(x) = 0$ is approximately π, so it follows that $\gamma_{mn} \approx \gamma_{1n} + (m-1)\pi$. Consequently the the commands

```
zeros =
Table[ FindRoot[ BesselJ[n,x] == 0,
                 {x, g1[[n]] + (m-1)Pi} ],
       {m,1,5}, {n,1,5} ];

g = x /. zeros;
g // TableForm

3.83171   5.13562   6.38016   7.58834   8.77148
7.01559   8.41724   9.76102   11.0647   12.3386
10.1735   11.6198   13.0152   14.3725   15.7002
13.3237   14.796    16.2235   17.616    18.9801
16.4706   17.9598   19.4094   20.8269   22.2178
```

yield a table displaying the *m*th zero γ_{mn} of $J_n(x) = 0$ in the *m*th row and *n*th column. For instance,

```
g[[2,3]]

9.76102
```

so we see that $\gamma_{23} \approx 9.76102$ is the 2nd zero of $J_3(x) = 0$.

Now that numerical values of the zeros of Bessel functions are available, we can employ the commands

```
x = r Cos[t];   y = r Sin[t];
m = 2;    n = 1;
ParametricPlot3D[
            {x,y,BesselJ[1, g[[m,n]] r] Cos[n t]},
             {r, 0,1}, {t, 0, 2Pi} ]
```

to display the initial position corresponding to the eigenfunction defined in (17). The next figure shows the cases $m = 2, n = 1$ and $m = 1, n = 2$.

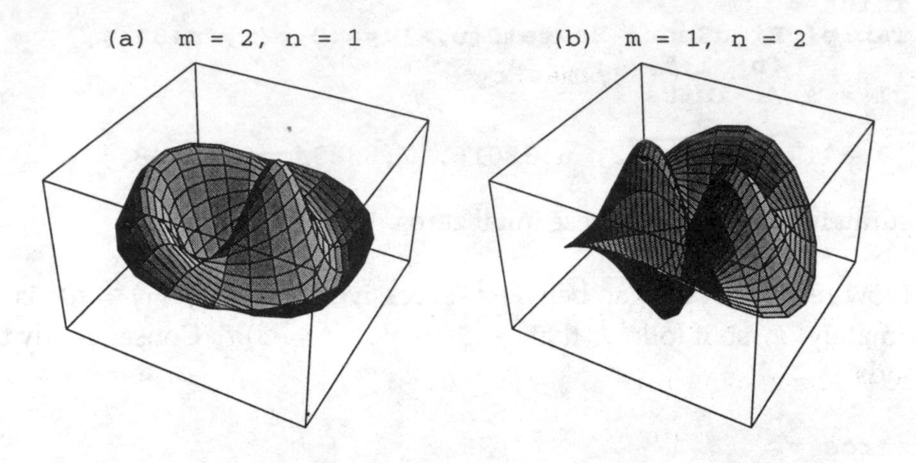

We suggest that you explore vibrating circular membrane possibilities by graphing convenient linear combinations of eigenfunctions defined as in (17). For instance, the figure below shows the initial position of the oscillation

$$u(r,\theta,t) = J_1(\gamma_{21}r)\cos\theta\cos\gamma_{21}t + J_2(\gamma_{32}r)\cos 2\theta\cos\gamma_{32}t \qquad (18)$$

defined for for a circular membrane with $a = 1$ and radius $c = 1$.

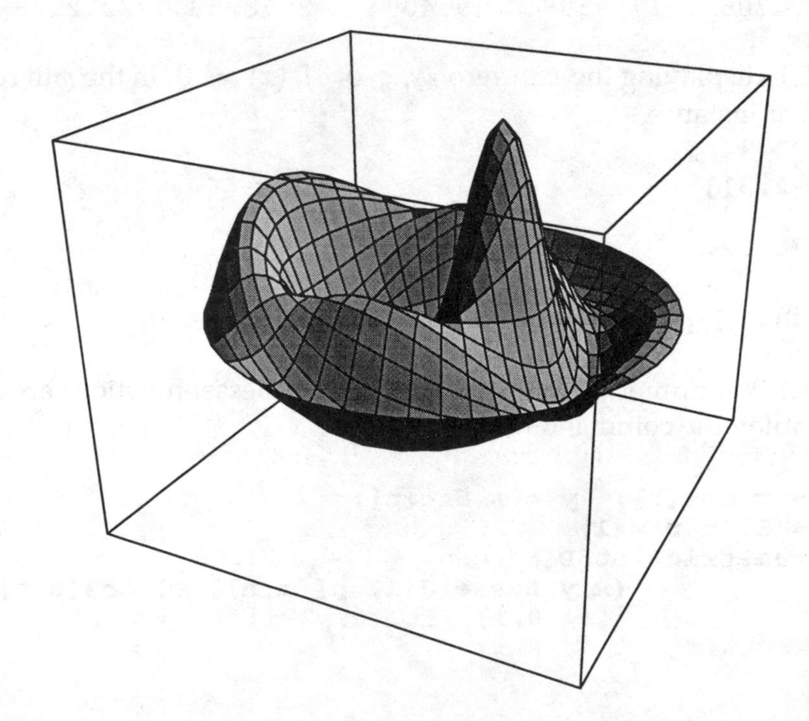

The following *Mathematica* commands generate a sequence of frames that can be animated to show a movie illustrating the circular membrane oscillation defined in (10).

```
k = 4;          (* cycles *)
p = 20;         (* frames/cycle *)

g12 = g[[1,2]];  g23 = g[[2,3]];
w1 = g12;        w2 = g23;          (* frequencies *)

x = r Cos[theta];  y = r Sin[theta];   (* polar coords *)

z = BesselJ[1, g12 r] Cos[  theta] Cos[w1 t] +
    BesselJ[2, g23 r] Cos[2 theta] Cos[w2 t];

Do[ ParametricPlot3D[ Evaluate[ {x,y,z} ],
                      {r, 0,1},    {theta, 0,2Pi} ],
    {t,0, k Pi/w1, Pi/(p w1)} ]
```

Project 46

Spherical Harmonic Waves

Reference: Sections 8.2 and 10.5 of Edwards & Penney
DIFFERENTIAL EQUATIONS with Computing and Modeling

In problems involving regions that enjoy spherical symmetry about the origin in space, it is appropriate to use spherical coordinates. The 3-dimensional Laplacian for a function $u(\rho,\phi,\theta)$ expressed in spherical coordinates is given by

$$\nabla^2 u = \frac{1}{\rho^2}\left[\frac{\partial}{\partial\rho}\left(\rho^2\frac{\partial u}{\partial\rho}\right)+\frac{1}{\sin\phi}\frac{\partial}{\partial\phi}\left(\sin\phi\frac{\partial u}{\partial\phi}\right)+\frac{1}{\sin^2\phi}\frac{\partial^2 u}{\partial\theta^2}\right]. \tag{1}$$

Note that $\rho = |OP|$ denotes the distance of the point P from the origin O, ϕ is the angle down from the vertical z-axis to OP, and θ is the ordinary polar coordinate angle in the horizontal xy-plane (though some texts reverse the roles of ϕ and θ). If u is independent of either ρ, ϕ, or θ then the corresponding second-derivative term is missing on the right-hand side in (1).

For instance, consider radial vibrations of the surface of a spherical planet of radius c. If $u(\phi,\theta,t)$ denotes the radial displacement at time t of the point (ϕ,θ) of the surface $\rho = c$ of the planet, then the wave equation $u_{tt} = a^2\nabla^2 u$ takes the form

$$\frac{\partial^2 u}{\partial t^2} = b^2 \nabla^2_{\phi\theta} u \tag{2}$$

where $b = a/c$ and

$$\nabla^2_{\phi\theta} u = \frac{1}{\sin\phi}\frac{\partial}{\partial\phi}\left(\sin\phi\frac{\partial u}{\partial\phi}\right) + \frac{1}{\sin^2\phi}\frac{\partial^2 u}{\partial\theta^2}. \tag{3}$$

Alternatively, Equation (2) models the oscillations of tidal waves on the surface of a spherical planet of radius c. In this case, $u(\phi,\theta,t)$ denotes the vertical displacement (from equilibrium) of the water surface at the point (ϕ,θ) at time t and $b^2 = gh/c^2$, where h is the average depth of the water and g denotes gravitational acceleration at the surface of the planet.

To solve equation (2) by separation of variables, show first that the substitution

$$u(\phi,\theta,t) = Y(\phi,\theta)\,T(t) \tag{4}$$

yields the equations

$$T'' + b^2\lambda T = 0, \tag{5}$$

$$\nabla^2_{\phi\theta} Y + \lambda Y = 0 \tag{6}$$

where λ is the usual separation constant. Next, show that the substitution

$$Y(\phi,\theta) = \Phi(\phi)\,\Theta(\theta) \tag{7}$$

in (6) yields the equations

$$\Theta'' + \mu\,\Theta = 0, \tag{8}$$

$$(\sin^2\phi)\,\Phi'' + (\sin\phi\cos\phi)\,\Phi' + (\lambda\sin^2\phi - \mu) = 0 \tag{9}$$

where μ is a second separation constant.

In order that $\Theta(\theta)$ be periodic with period 2π, it follow from (8) that $\mu = m^2$, the square of a non-negative integer, in which case a typical solution of (8) is

$$\Theta_m(\theta) = \cos m\theta. \tag{10}$$

Now show that with $\mu = m^2$ the substitution

$$x = \cos\phi \tag{11}$$

in (9) yields the ordinary differential equation

$$\left(1-x^2\right)\frac{d^2\Phi}{dx^2} - 2x\frac{d\Phi}{dx} + \left(\lambda - \frac{m^2}{1-x^2}\right)\Phi = 0. \tag{12}$$

Note that if $m > 0$ then (12) is a Legendre equation with dependent variable Φ and independent variable x. According to Section 8.2, this equation has a solution $\Phi(x)$ that is continuous for $-1 \le x \le 1$ provided that the parameter $\lambda = n(n+1)$ where n is a non-negative integer. In this case the continuous solution $\Phi(x)$ is a constant multiple of the nth degree Legendre polynomial $P_n(x)$.

Equation (12) is an **associated Legendre equation**, and it likewise has a solution $\Phi(x)$ that is continuous for $-1 \le x \le 1$ provided that the parameter $\lambda = n(n+1)$ where n is a non-negative integer. In this case the continuous solution $\Phi(x)$ is a constant multiple of the **associated Legendre polynomial**

$$P_n^m(x) = (1-x^2)^{m/2}\, P_n^{(m)}(x), \tag{13}$$

where the mth derivative of the ordinary Legendre polynomial $P_n(x)$ appears on the right. For instance, writing **Pmn** for $P_n^m(x)$, *Mathematica* gives

```
P00 = LegendreP[0,x]
1

P01 = LegendreP[1,x]
P11 = Sqrt[1-x^2] D[LegendreP[1,x],x]
x
                2
Sqrt[1 - x ]

P02 = LegendreP[2,x]
P12 = Sqrt[1-x^2] D[LegendreP[2,x],x]
P22 = (1-x^2) D[LegendreP[2,x],{x,2}]
            2
-1 + 3 x
---------
    2

                    2
3 x Sqrt[1 - x ]
            2
3 (1 - x )
```

Actually, the associated Legendre functions are built into *Mathematica*, with **LegendreP[n,m,x]** denoting $(-1)^m P_n^m(x)$. (Many references include *Mathematica*'s sign $(-1)^m$ in the definition of $P_n^m(x)$.)

Using *Maple*, you must first load the orthogonal polynomials package. Only the ordinary Legendre functions are immediately available, so you must implement the definition in (13).

```
> with(orthopoly):
> p4 := P(4,x);
```

$$p4 := \frac{35}{8}x^4 - \frac{15}{4}x^2 + \frac{3}{8}$$

```
> p24 := expand((1-x^2)*diff( p4, x$2));
```

$$p24 := 60\,x^2 - \frac{15}{2} - \frac{105}{2}x^4$$

If **x** is a row vector with k elements then the MATLAB command **legendre(n,x)** yields an $(n+1)\times k$ matrix whose mth row contains the values of P_n^{m-1} at the elements of **x**. Thus, the computation

```
» legendre(3, 0:1/3:1)
ans =
          0   -0.4074   -0.2593    1.0000
    -1.5000   -0.6285    1.3665         0
          0    4.4444    5.5556         0
    15.0000   12.5708    6.2113         0
```

shows that $P_3^2(1/3) \approx 4.4444$ and $P_3^2(2/3) \approx 5.5556$. (It would be instructive for you to deduce from (13) the exact values $P_3^2(1/3) \approx \frac{40}{9}$ and $P_3^2(2/3) \approx \frac{50}{9}$.

At any rate, given non-negative integers m and n with $m \le n$, substitution of $x = \cos\phi$ in the continuous solution (13) of Eq. (12) with $\lambda = n(n+1)$ yields the solution

$$\Phi_{mn}(\phi) = P_n^m(\cos\phi) = (\sin\phi)^m P_n^{(m)}(\cos\phi) \tag{14}$$

of Eq. (9) with $\mu = m^2$ and $\lambda = n(n+1)$. Substitution of $\lambda = n(n+1)$ in (5) yields the typical solution

$$T_n(t) = \cos \omega_n t \tag{15}$$

with frequency

$$\omega_n = b\sqrt{n(n+1)}. \tag{16}$$

Putting it all together, we get finally the eigenfunction

$$u_{mn}(\phi,\theta,t) = P_n^m(\cos \phi) \cos m\theta \cos \omega_n t \tag{17}$$

($0 \le m \le n = 1, 2, 3, \ldots$) of the wave equation in (2). The remaining eigenfunctions are obtained by (independently) replacing $\cos m\theta$ with $\sin m\theta$ and $\cos \omega_n t$ with $\sin \omega_n t$.

With all this preparation, your task is to investigate graphically the way in which water waves slosh about on the surface of our small planet. Let us take a sphere of radius $c = 5$ with $b = 1$ in (2), and (somewhat unrealistically) consider waves of amplitude $h = 2$. The following MATLAB function **spharm(m,n)** constructs a $\phi\theta$–grid on the surface of the sphere and calculates the corresponding matrix Y of values of the surface spherical harmonic function $Y_{mn}(\phi,\theta) = P_n^m(\cos \phi) \cos m\theta$.

```
function    [Y,phi,theta] = spharm(m,n)

% constructs the mn_th spherical surface harmonic

phi =    0 : pi/40 : pi;          % co-latitude
theta = 0 : pi/20 : 2*pi;   % polar coord angle

[theta,phi] = meshgrid(theta,phi);
Theta = cos(m*theta);

Phi = legendre(n, cos(phi(:,1)));
Phi = Phi(m + 1,:)';
pp = Phi;
for k = 2 : size(Phi,1)
    Phi = [Phi pp];
      end;
Y = Phi.*Theta;
m = max(max(abs(Y)));
Y = Y/m;
```

The function **spshape(m,n)** then displays the corresponding initial graph $\rho = c + h\,Y_{mn}(\phi,\theta)$. The next figure show such plots for $n = 6$ and $m =$ 0, 1, 2, 3, 4 generated by commands such as

```
» spshape(1,6)
```

```
function    spshape(m,n)

% Displays shape of mn_th eigenfunction for waves
% of height h on surface of planet of radius c.

c = 5;     h = 2;

[Y,phi,theta] = spharm(m,n);
rho = c + h*Y;
r = rho.*sin(phi);
x = r.*cos(theta);
y = r.*sin(theta);
z = rho.*cos(phi);

mesh(x,y,z)
axis('square')
axis([-6 6 -6 6 -6 6])
axis('off')
view(40,30)
colormap(white)
```

m = 0, n = 6

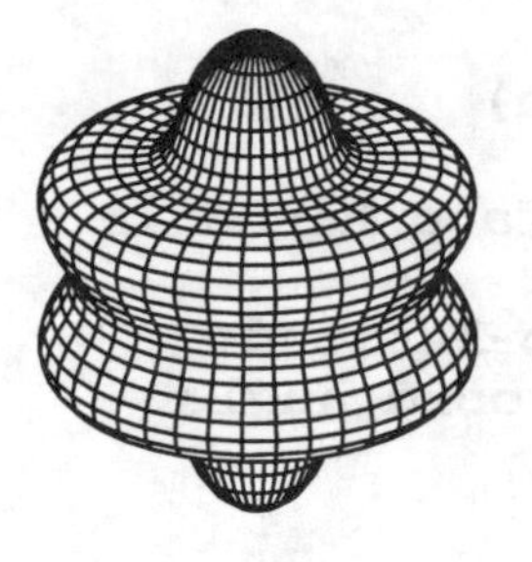

m = 1, n = 6

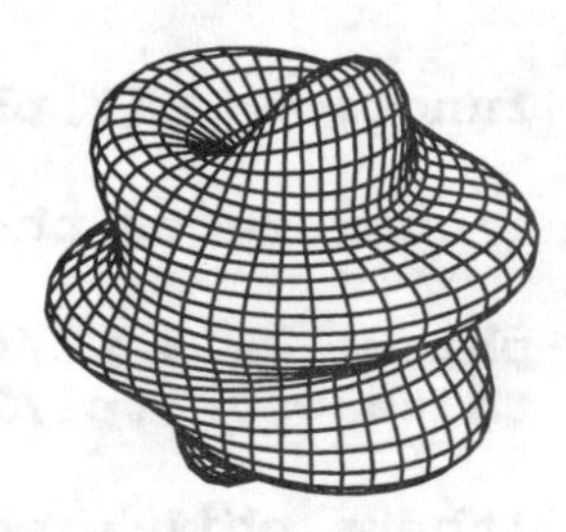

m = 2, n = 6

m = 3, n = 6

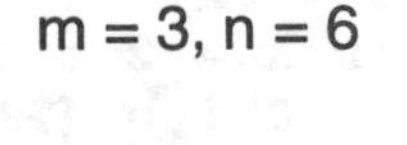

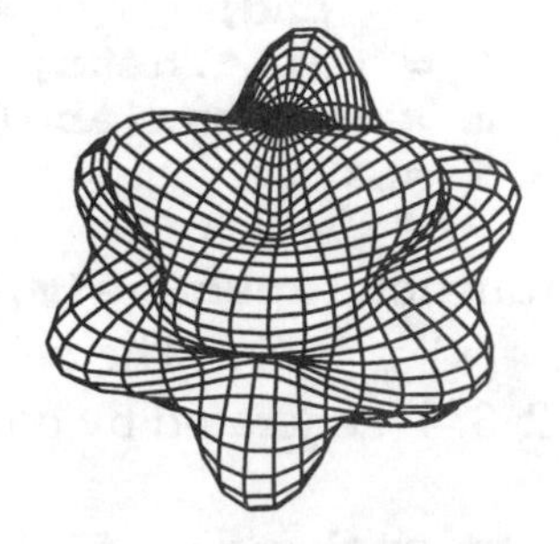

```
function  Mslosh = spmovie(m,n)

% Constructs a movie showing a vibration of
% spherical corresponding to the m,n-eigenfunction.

c = 5;    h = 2;

w = sqrt(n*(n+1));
k = 20;                                    % steps per cycle
dt = 2*pi/(k*w);                           % (time) step size

[Y,phi,theta] = spharm(m,n);

Mslosh = moviein(k);
for j = 0 : k-1
    t = j*dt;
    rho = c + h*Y*cos(w*t);
    r = rho.*sin(phi);
    x = r.*cos(theta);
    y = r.*sin(theta);
    z = rho.*cos(phi);
    surf(x,y,z)
    axis('square')
    axis([-6 6 -6 6 -6 6])
    axis('off')
    view(40,30)
    colormap(jet)
    Mslosh(:,j+1) = getframe;
    end

movie(Mslosh,5)
```

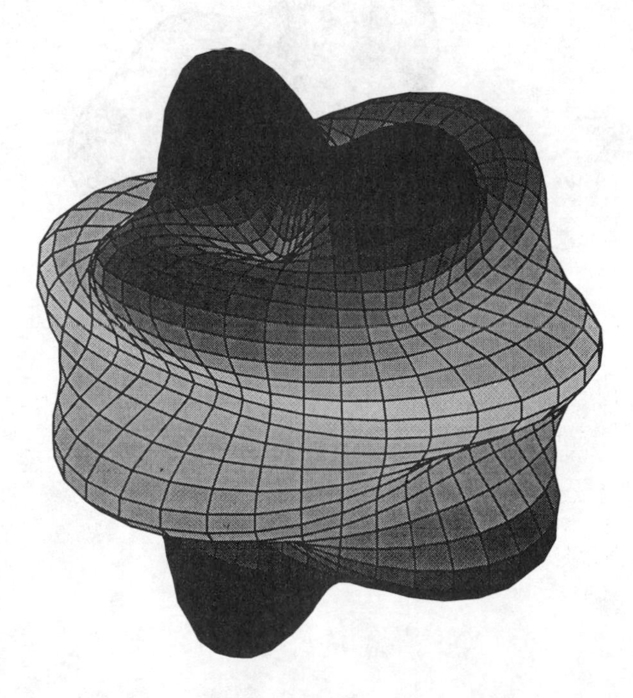

The function **spmovie(m,n)** constructs and displays a movie consisting of $k = 20$ frames per complete oscillation. The figure on the preceding page shows the initial frame of the movie generated by the command

```
» Mslosh = spmovie(2,7);
```

The "movie matrix" **Mslosh** contains over a quarter-million elements and occupies over 2 megabytes of memory. With a Macintosh system it can be saved as a QuickTime movie with the command

```
» qtwrite(Mslosh, jet, 'movie27');
```

Construct some movies of your own. If you're ambitious you can investigate linear combinations of different spherical surface harmonics. For instance, the figure below shows the initial position of a *Mathematica*-generated movie showing the oscillation with initial position function

$$u(\phi,\theta,0) \;=\; 5+2\,P_4^1(\cos\phi)\cos\theta+P_6^3(\cos\phi)\cos 3\theta.$$

Does the result (in full motion and living color) remind you (at least vaguely) of a throbbing, beating human heart?

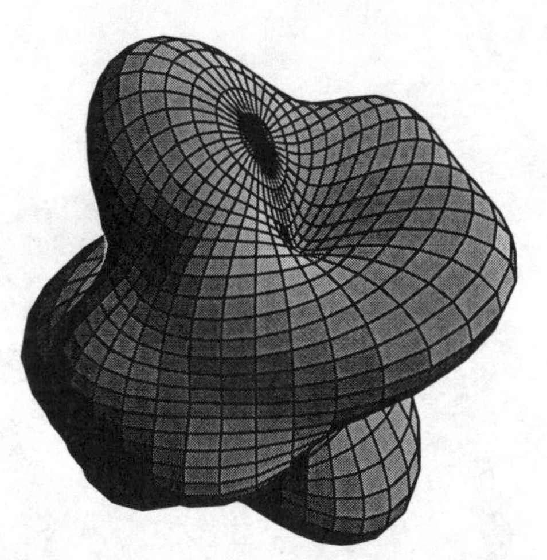